专家编写服装实用教材：中级版

服装基础英语（第 5 版）

田守华　编著

中国纺织出版社有限公司

内 容 提 要

本书对第4版进行了修改补充，删去原教材中过时的材料和部分练习，增加了大量服装专业文章，同时也增加了新知识、新工艺。本书注重讲解服装基础英语知识，注重对服装专业英语表达方法的学习和操练，包括了大量服装专业英语课文和阅读材料，对重点知识、术语、专业词汇做到反复出现、反复练习，以达到熟练掌握的目的。书中内容涉及服装制作工具、服装材料、选料、量体、样板制作、排料、裁剪、缝纫工艺、熨烫、包装等，并且增加了服装跟单、对外贸易、业务洽谈、签订合同等方面的知识，使教材内容更新、涵盖面更宽，更有利于学生专业能力的培养和英语专业水平的提高。

本书适用于服装专业师生参考学习，也可作为时尚爱好者对于相关知识的学习拓展。

图书在版编目（CIP）数据

服装基础英语 / 田守华编著 . --5 版 . -- 北京：中国纺织出版社有限公司，2022.1

专家编写服装实用教材：中级版

ISBN 978-7-5180-9061-7

Ⅰ. ①服… Ⅱ. ①田… Ⅲ. ①服装工业—英语—高等学校—教材 Ⅳ. ①TS941

中国版本图书馆 CIP 数据核字（2021）第 217503 号

责任编辑：孙成成　施　琦　　责任校对：王花妮

责任印制：王艳丽

中国纺织出版社有限公司出版发行

地址：北京市朝阳区百子湾东里 A407 号楼　邮政编码：100124

销售电话：010—67004422　传真：010—87155801

http://www.c-textilep.com

中国纺织出版社天猫旗舰店

官方微博 http://weibo.com/2119887771

三河市宏盛印务有限公司印刷　各地新华书店经销

1998 年 8 月第 1 版　1999 年 12 月第 2 版　2009 年 12 月第 3 版

2014 年 7 月第 4 版　2022 年 1 月第 5 版第 1 次印刷

开本：787 × 1092　1/16　印张：16

字数：324 千字　定价：55.00 元

前言
Preface

为满足服装专业教学需要，为进一步提高学生们的英语阅读能力和专业英语应用能力，我们对原教材进行了修改补充，删去原教材中过时的材料和部分练习，增加了大量服装专业文章，同时也增加了新知识、新工艺。本书注重讲解服装专业英语知识，注重对服装专业英语表达方法的学习和操练，包括了大量服装专业英语课文和阅读材料，对重点知识、专业术语、专业词汇做到反复出现、反复练习，以达到熟练掌握的目的。本教材内容涉及服装制作工具、服装材料、选料、量体、样板制作、排料、裁剪、缝纫工艺、熨烫、包装等，并且增加了服装跟单、对外贸易、业务洽谈、签订合同等方面的知识，使内容更新、涵盖面更宽，更有利于学生专业能力的培养和英语专业水平的提高。

黄骏行老师参加了本书第1版的部分编写，在此表示衷心感谢。《服装基础英语》一书的编写得到了邹建明、田霭俊、许振华、王文侠、朱理、徐礼莱、李德宝、李修珍、周璟等同志的帮助，在此表示感谢！

希望本教材修改后能受到广大师生和读者的欢迎。对本教材不足之处恳请读者批评指正。

编著者

2021年8月

Contents

目录

Lesson One

Text 1

>> Sewing Machines 缝纫机

There are a great many sewing machines in our factory. We know how to handle the machines. They are very important in our production. On the worktable there is a sewing machine, a bobbin and a bobbin case. This is a sewing machine. A sewing machine is an instrument which you use to sew pieces of cloth together. It is a mechanical device for sewing fabric and other materials. These sewing machines are all made in our country. Those sewing machines are automatically controlled. The sewing machines are operated easily. We have many kinds of sewing machines now. Some of them can be used for sewing fabrics and pieces of clothes, some of them can be used for buttonholes, and some of them can be used for sewing buttons on clothes, and some sewing machines for embroidery. They work well. This is a sewing machine body. Those are sewing machine parts and accessories. These parts and accessories are small but necessary and important. The number of every sewing machine parts is the same. All these sewing machine parts are replaceable. This is a bobbin. A bobbin is a small spool on which thread is wound. Here is a bobbin case. We need many sewing machine bobbins in operation. That is a needle. It is a sewing machine needle. It is used to join pieces of cloth with thread. Sewing machines are very important and useful instruments in our work. The sewing machine is a must for garment workers. We use sewing machines to make up garments and other things. We operate sewing machines every day. The operation of a sewing machine is easy. Sewing machines give us a lot of help in our work. Some sewing machines have been introduced to simplify the work. When operating sewing machines, we must pay special attention to safety. Sewing machines are our good friends. They work smoothly. A good sewing machine can perform more efficiently than a large group of people.

New Words and Expressions

sewing machine	缝纫机	bobbin	梭芯
a great many	许多	spool	线轴
handle	操纵	thread	线
production	生产	wind	缠绕
worktable	工作台	bobbin case	梭壳
instrument	器具	in operation	操作中
sew...together	将……缝在一起	be used to do	被用作
pieces of cloth	衣片	needle	针
mechanical device	机械装置	a must	不可缺少的东西
automatically	自动地	operation	操作
control	控制	simplify	简化
operate	操作	sewing machine needle	缝纫机针
embroidery	绣花	join	结合
replaceable	可替换的	efficiently	有效地
fabric	布料	smoothly	顺利地
buttonhole	纽洞	sew...on	将……缝上
button	纽扣	accessory	配件
sewing machine body	机身	special	特别的
make up garments	缝制服装	attention	注意
perform	履行，执行	safety	安全

Notes to the Text 1

1. A sewing machine is an instrument which you use to sew pieces of cloth together. 缝纫机是一种将布料缝在一起的机械。“sew pieces of cloth together”将布片缝在一起。例如：Let's sew two pieces of cloth together. She sewed two pieces of cloth together.

2. ...mechanical device 机械装置。例如：We invented a safety mechanical device to repair the machine. 我们发明了一种安全机械装置修理这台机器。

3. These sewing machines are automatically controlled. Those sewing machines are operated easily. 这些缝纫机都是自动控制的。那些缝纫机操作起来很容易。

4. They work well. 它们运转顺利。

5. The number of every sewing machine parts is the same. 每台缝纫机的零件都是相同的。“the number of”意思是“……的数量”，谓语动词用单数。例如：The number of the jackets is 22. “a number of”意思是“许多”，修饰主语时谓语动词用复数。例如：A number of new dresses are on display at the fair.

6. These parts and accessories are small but necessary and important. 这些零件和附件虽然小但却是必要的、重要的。

7. A bobbin is a small spool on which thread is wound. 梭芯是绕线的小线轴。

8. The sewing machine is a must for garment workers. 缝纫机是服装工人的必需品。“must”作名词是“必须的东西”。例如：Warm clothes are a must in winter.

9. operate a sewing machine = work a sewing machine 操作缝纫机。

10. Some sewing machines have been introduced to simplify the work. 许多缝纫机引进后简化了工作。

11. sewing machine parts and accessories 缝纫机零配件。

12. A good machine can perform more efficiently than a large group of people. 一台好缝纫机的效率超过许多人。

Text 2

>> Our Sewing Machines 我们的缝纫机

Here we have a sewing machine on the worktable. The sewing machine is a must for us. It is an important instrument in our tailoring and dressmaking. It is necessary for us to have practical and useful instruments in production of high quality clothes. Sewing machines are essential to our production. Sewing machines are common and convenient instruments in production. There are many kinds of sewing machines in our factory. Each has its own usage and function. A sewing machine is made up of different parts. They are bobbins, bobbin cases, screws and accessories. Every kind of sewing machine is very useful in our production. Every sewing machine must be in good condition. With the help of these sewing machines, we produce high quality garments. We have these sewing machines for sewing fabrics and pieces of clothes. We have those sewing machines for buttonholes and stitching buttons on clothes. Look, this is a sewing machine needle. It is pointed and sharp.

We can make good work with sewing machines. We must clean and oil the sewing machines every day. We must keep the sewing machines working well. We ensure the sewing machines perform well every day during working hours. Sewing machines in good conditions is very important. Sewing machines give us a lot of help. Every day we must keep our sewing machines in good conditions. A good sewing machine can perform more efficiently than a large group of people.

New Words and Expressions

essential	重要的	quality	质量
practical	实用的	different parts	各种零件
useful	有用的	stitch	缝上
high quality	优质的	stitch...on...	将……缝在……上
common	普通的	sharp	锋利的
in good condition	状况良好	oil the machine	给缝纫机加油
convenient	方便的	with the help of	借助于
usage and function	用途和功能	keep...working well	使……运转顺利
be made up of	由……组成	ensure	保证
working hours	工作期间	perform well	运转正常
group	队，组	more efficiently than	比……更有效
in production	生产中	pointed	尖的

Notes to the Text 2

1. The sewing machine is a must for us. 缝纫机对我们来说是不可缺少的设备。这

里“must”是名词，意思是“不可缺少的东西”。例如：The book is a must for beginners of clothes. 这本书对服装初学者是必不可少的。

2. It is necessary for us to have practical and useful instruments in production of high quality clothes. 对我们来说要生产优质的服装拥有合适实用的缝纫机是必要的。

3. With the help of the these sewing machines, we produce high quality garments. 凭借着这些缝纫机，我们生产优质服装。“with the help of...”意思是“凭借着，借助于，在……的帮助下”。例如：With the help of the dictionary, I can read the book. 借助于字典，我能看懂这本书。With the help of the teacher, I draw the patterns. 在老师的帮助下，我绘制了这些纸样。

4. We must keep the sewing machines working well. 我们必须保持缝纫机运行良好。

5. Every day we must keep our sewing machines in good conditions. 我们必须保证缝纫机每天都处于良好的状态。

Substitution Drills 替换练习

1. These are
   - bobbin cases
   - bobbins
   - sewing machine parts

   .

2. Those are
   - sewing machine bodies
   - sewing machine parts and accessories

   .

3. These
   - sewing machines
   - computers
   - steam irons

   are all made in our country.

4.
   - Sewing machines
   - Computers
   - Steam irons

   are very important and useful instruments in our work.

5. We use sewing machines / needles / bobbins every day.

Exercises to the Texts 课文练习

1. 反复朗读并抄写本课单词
2. 牢记本课词组和句型
3. 翻译下列词组

1）bobbin case
2）sewing machine needle
3）simplify our work
4）sew pieces of cloth together
5）automatically controlled
6）be operated easily
7）parts and accessory
8）the number of
9）a number of
10）a must
11）in operation
12）important and useful instrument
13）make up a jacket
14）wind the thread on the bobbin
15）a large group of students
16）with thread
17）wind thread on a spool
18）a lot of help
19）work smoothly
20）a great many

4. 将下列短语译成英语

1）将……缝在一起
2）衣片
3）缝纫机针
4）棉线
5）比……更有效
6）运转顺利
7）自动控制
8）操作缝纫机
9）优质
10）由……组成
11）梭壳
12）一块料子
13）简化工作
14）梭芯
15）工作期间
16）状况良好
17）运转正常
18）在……生产过程
19）借助于
20）实用的

5. 翻译

1）我们有很多种缝纫机零配件。
2）他正在往梭芯上绕线。
3）这些缝纫机都是中国制造的。
4）缝纫机在我们工作中很重要。

5）缝纫机是他的必需品。
6）确保缝纫机运转顺利。
7）这根针又尖又锋利。
8）借助字典她读了今天的报纸。
9）有一台缝纫机对我来说是必要的。
10）操作缝纫机是不容易的。
11）那些是缝纫机针和手工针。
12）她正在缝一块料子。
13）缝纫机给我们的工作提供了很多帮助。
14）有些缝纫机引进后简化了工作。
15）我们这里有缝纫机零配件。
16）这些梭壳是可替换的。
17）我们班的学生人数是48人。
18）这个车间有很多缝纫机。
19）我们必须保持缝纫机干净。
20）那些机器的运行状况很好。

Conversation

>> Sewing Machines and Needles 缝纫机和针

Conversation 1

A: Oh, look, there are so many strange needles here, they are different sizes, what are these needles for?

B: They are different needles. These are hand needles, some people call them sewing needles. Those are sewing machine needles. We need many different needles for daily production. They are fit for various uses. People used to do their sewing work with hand needles. Now most people do it with sewing machines.

A: What's the difference between hand needles and sewing machine needles?

B: Hand needles are used for needlework or manual sewing while sewing machine needles are on the sewing machines used in sewing machine work. Their usages are different and their shapes are different.

A: I see. They are both sharp and pointed.

B: Yes, we mustn't leave any hole in the garments when we sew and stitch garments, so they are very important tools in making up garments, and we must take care of them when we are using them in our sewing work. We mustn't leave any needles in the garments any time. It's dangerous.

Conversation 2

A: It is a sewing machine, isn't it?

B: Yes. This is the sewing machine body. That is a bobbin case. These are bobbins. Bobbin are our daily uses. We need many bobbins in production. Different colors of thread need different bobbins.

A: What's the usage of a sewing machine?

B: We use sewing machine to make up garments and other things. Sewing machine is a very important tool in our production.Modern production would be impossible without sewing machines.

A: It is great fun making up garments.

B: Yes. It's interesting. It gives me great pleasure to make up a garment. But accidents usually happen to those who have neglected to observe the rules of work. The problem is too serious to be neglected. So we must pay attention to safety. Safety first!

A: I'll remember it.

New Words and Expressions

hand needle	手工针
machine needle	缝纫机针
needlework	手工活
used to do	过去常常
difference	区别
be fit for...uses	适合……用途
shape	形状
make up	缝制
manual sewing	手工缝纫活
happen	发生
I see	我明白了
be used for	被用作
fun	有趣
those who	凡是……的人
neglect	忽视
observe	遵守
give sb. great pleasure to do sth.	做某事给了某人很大乐趣
serious	严重的
It's great fun doing...	做……有趣
leave	留下

Reading Practice 1

Tools and Sewing Machine Parts 工具和缝纫机零部件

We have our needlework lesson now. Here we have many needles. The needles are our daily uses. They can be divided into two kinds, sewing machine needles and hand needles or sewing needles. This is a hand needle. We can also say it is a sewing needle. Now let's thread a needle. That is a sewing machine needle. They are needles. These are hand needles and those are sewing machine needles. They are very nice. They are sharp and pointed. The eye of a needle is very small. It's not easy to thread. Now let's pass a thread through the eye of a needle. Look, the needle eye is too small to hold the thread. Let's do it a second time. We take the hand needles and thread, and do a bit of needlework. We sew a button with a needle and cotton. We sew buttons on a coat. These are thimbles. Those are rulers. Here is a metal thimble. That is a wooden ruler. It is a long ruler. Here is a tape-measure. Needles, thimbles tape-measures and rulers are useful tools in our needlework. We should be careful to use the tools and take good care of the tools.

These are bobbins. Those are bobbin cases. Here are some sewing machine screws. These are sewing machine gears. They are sewing machine parts. There are many parts in a sewing machine. We have many spare parts there. They are very important to sewing machines. We use needles, thimbles, rulers and sewing machines every day. They are very important. We must take care of them. We should clean them every day. If you take good care of your sewing machines, tools and needles, you'll increase their lives.

New Words and Expressions

thread	给……穿线
thread a needle	给针穿线
thimble	顶针
screw	螺丝
gear	齿轮
take care of	照顾
a second time	再来一次
spare parts	备件
pass the thread through	将线穿过
hold	穿过
tape-measure	软尺
too...to	太……以至于不
increase	延长

Reading Practice ❷

Needles and Sewing Machine Parts 针和缝纫机零件

Here we have some needles. They are different in sizes and usages. Some of them are hand needles and some of them are sewing machine needles. A hand needle is a small, thin of steel pointed at one end and with a small hole at the other end for thread. We sew cloth with hand needles in our needlework lesson. We do needle work with a hand needle and thread. We can sew two pieces of cloth together with a hand needle.

A sewing machine needle is a small, thin of steel pointed with a small hole for thread at one end and with flat end at the other end. A sewing machine needle is very useful. We can sew pieces of cloth together with a sewing machine. The quickest way to sew is with a sewing machine. Look, there are many sewing machines over there. They are all made in our country.

Let's look at these sewing machine parts, bobbins, bobbin cases, screws and gears. We need many bobbins and bobbin cases in our tailoring and dressmaking. There are many screws and gears in a sewing machine. Every screw and gear has its own place. Many sewing machine parts are kept in place by screws. We must know the importance of these sewing machine parts.

We have many useful tools in our tailoring and dressmaking. They are thimbles, rulers and tape-measures. They are of great help to our work. So we must take care of them.

New Words and Expressions

different in size	不同尺寸
steel pointed	钢尖头
sew two pieces of cloth together	将两块布缝在一起
flat end	平头
pieces of cloth	几块布
the quickest way to do sth	做某事最快的方法
one end	一头
the other end	另一头
sewing machine parts	缝纫机零件
screws and gears	螺丝和齿轮
have its own place	在自己的位置
the importance of	……的重要性
be of great help to	对……有很大帮助
take care of	照料

Lesson Two

Text 1

>> Tailors, Dressmakers, Cutters and Designers
男装师、女装师、裁剪师和设计师

I am a tailor. I work in a garment company. A tailor is a person who makes clothes for people. People usually say that a tailor is a garment worker. We are tailors in our company. They are dressmakers. A dressmaker is a person who makes women's dresses. The difference between a tailor and a dressmaker is what clothes he or she makes up. If he makes clothes for men, usually he is a tailor. If he makes women's dresses, he is a dressmaker. We are all workers in our company. We engage in different but related work. Everyone should do his job well. We tailors and dressmakers work together every day.

They are cutters. A cutter is a person who cuts out suits, dresses and other clothes. We tailors and dressmakers work in the sewing workshops. The cutters work in the cutting workshops. The cutters cut out the pieces according to the patterns. They pile up the pieces and number all the pieces of clothes. We are all skilled workers. We are all qualified garment workers.

Tom and Lucy are designers. A designer is a person who designs clothes or dresses for men, women, children and other people. We introduced CAD and computers in our design. A computer can complete the work in a few minutes, which used to take us a few days to do. We have many creative ideas. Their creative ideas makes the work very fruitful. The designers in our company design beautiful clothes every year. The products of our designers design are more interesting than those of others. We win many prizes every year. Our products have enjoyed wide popularity in the market. Everybody in our company loves them. We tailors, dressmakers, cutters and designers are all garment workers. We work in a company and work together happily. We do our work with pleasure.

We have a lot of sewing machines and other machines in our workshops. Many new machines have been introduced recently in the workshops. These new machines have raised efficiency many times. We have made much improvement in the quality of our products. We put emphasis again and again on the importance of the quality.

>> New Words and Expressions <<

tailor	裁缝，服装工人
dressmaker	女装师
cutter	裁剪工
designer	设计师
work together	一起工作
skilled	有技术的
creative	创新的
recently	最近
design	设计
efficiency	效率
raise	提高
and others	等等，其他人
quality	质量
emphasis	强调
put emphasis on	强调
number	给……编号
pile up	堆起
qualified	合格的
fruitful	有成果的
workshop	车间
company	公司
prize	奖
CAD	计算机辅助设计
complete	完成
take.sb...to do	花……（时间）做某事
happily	幸福地
fully	完全地
product	产品
enjoy wide popularity	享有很高的声誉
improvement	改进
again and again	再三
importance	重要性
engage in	从事

Notes to the Text 1

1. The difference between a tailor and a dressmaker is what clothes he or she makes up. “A tailor” 和 “A dressmaker” 的区别是他或她做的是什么衣服。

2. We are all qualified garment workers. 我们是完全合格的服装工人。

3. We engage in different but related work. 我们从事着不同但相关的工作。

4. Many new machines have been introduced recently in the workshops. 这车间最近引进了许多新机器。

5. The products of our designers design are more interesting than those of others. 我们设计师设计的产品比其他设计师的产品更有趣。

6. A computer can complete the work in a few minutes, which used to take us a few days to do. 一台计算机能在几分钟内完成过去要几个人花几天才能做完的事。

7. These new machines have raised efficiency many times. 这些新机器将效率提高了好几倍。“many times” 是 “好几倍，好几次” 的意思。

8. We do our work with pleasure. 我们工作（干)得很愉快。“do sth with pleasure” 是 “某事做得很愉快” 的意思。

9. Our products have enjoyed wide popularity in the market. 我们的产品在市场上享有很高的声誉。“enjoy wide popularity” 意思是 “享有很高的声誉”。例如：Our garments enjoy wide popularity in their country.

10. We put emphasis again and again on the importance of the quality. 我们再三强调质量的重要性。“put emphasis again and again on...” 意思是 “再三强调”。例如：The teachers put emphasis again and again on the importance of learning English well.

Text ❷

>> They Are Tailors, Dressmakers, Cutters and Designers of our Factory
他们是我厂的男装师、女装师、裁剪师和设计师

They are tailors of our factory. They are all master workers. They stitch clothes such as shirts, suits, jackets, coats and overcoats for men every day. Look, they are sewing and stitching garments, the suits over there are well tailored. Whether the workers can do a good job depends on their skills. Every worker is a very efficient person and he usually finishes his work ahead of schedule. A worker's ability will improve with practice. Everybody is improving in skill. The fine workmanship and good quality of our garments are popular in many markets.

Now let's go into another workshop. The workers here are all dressmakers. They stitch dresses, skirts, suits and blouses for women. Sometimes people call them seamstresses. The clothes here are all fashionable. They make the fashionable dresses, skirts and blouses. They produce thousands of pieces of fashion every year. They are the fashion of the day. They have won reputation for our company. Our products have become best sellers in the market. Our products are very popular with customers. This has become the new trend in women's apparel.

Here we are in the cutting workshop. You can see the machines here are different from the ones in sewing workshop. The workers here are cutters. They are running the machines. The workers are cutting out suits, dresses, coats and other clothes. They are skilled workers. Some machines here are introduced recently, and some of them are invented by the workers and engineers. Much improvement has been made in the safety. These machines have raised efficiency many times.

Here comes a designer. A designer is a person whose job is to design clothes for men, women and children. The designers in our factory have much talent for clothes. They have creative ideas and turn the creative ideas into production. With the help of computers they design all kinds of suits, coats, jackets, blouses and trousers for customers. They win many prizes every year.

New Words and Expressions

master	师傅
overcoat	大衣
be well tailored	做工精良的
go into	进入
seamstresses	缝纫女工
fashionable	时髦的
fashion of the day	目前流行的时装
trend	时尚，潮流
apparel	服装
run the machine	操作运转机器
introduce	引进
much improvement	很大改进
workmanship	工艺
be popular in	流行
raise efficiency	提高效率
have talent for	在……有才能
win prize	获奖

Notes to the Text 2

1. Every worker is a very efficient person and he usually finishes his work ahead of schedule. 每个工人都是做事很有效率的人，通常都会提前完成工作。

2. Much improvement has been made in the safety. 安全性方面得到了很大的改进。

3. ...the new trend in women's apparel. 女装的新时尚。

4. Here comes a designer. 一位设计师朝这儿走来。这是个倒装句，以“here, there”开头的句子，主语如果是名词用倒装形式，主语如果是代词用顺装形式。例如：Here comes the bus. Here comes a teacher. Here you are.给（你）。Here we are. 我们到了。Here it is.（东西）在这儿。

5. They have creative ideas and turn the creative ideas into production. 他们有创新性的想法并将这些想法运用到生产中。

6. with the help of sb. 在某人的帮助下。例如：With the help of the master, he repaired the sewing machine. 在师傅的帮助下他修好了这台缝纫机。with the help of sth. 借助于某物（事）。With the help of the ruler, he made the paper patterns. 借助于这把尺，他制作了这些纸样。

7. The fine workmanship and good quality of our garments are popular in many markets. 我们服装的精湛工艺和优异的质量在市场上很受欢迎。

Substitution Drills 替换练习

1. We are [tailors / dressmakers / cutters / designers] in our factory.

2. What's the difference between [a tailor and a dressmaker / A and B / this machine and that one] ?

3. It takes [us two hours / him three days / them four weeks / me five minutes] to finish it.

4. [The new machines / The new computers / The new ways] have raised efficiency many times.

5. Many [new machines / sewing machines / new ways] have been introduced recently in the workshops.

6. We work in the [sewing / cutting / pressing] workshop.

7. The designers in our company designed many [dresses / jackets / suits] .

8. Tom and Helen are [tailors / cutters / dressmakers] .

Exercises to the Texts 课文练习

1. 牢记本课词组和句型
2. 翻译下列短语

1）used to do
2）the difference between A and B
3）win a prize
4）skilled dressmaker
5）stitch clothes
6）such as
7）be well tailored
8）fine workmanship
9）be popular in
10）go into
11）complete the work
12）introduce CAD in
13）take us an hour to do
14）go to work together
15）the fashion of the day
16）new trend
17）women's apparel
18）much improvement
19）raise efficiency
20）Here you are.

3. 将下列词组译成英语

1）设计漂亮的衣服
2）将Mary介绍给Tom
3）给孩子们做衣服
4）一起学习
5）为……设计服装
6）获奖
7）很有才能
8）借助于
9）车来了
10）手里那本书
11）一个努力工作的人
12）设计衬衫
13）某人做某事花……时间
14）提高效率
15）操作缝纫机
16）干得愉快
17）被用作
18）过去常
19）完成工作
20）工人师傅

4. 汉译英

1）你的衬衣和他的衬衣有什么区别?
2）有技能的女装师能很容易地完成这件衣服。
3）我们每年引进新式女装。
4）设计这件连衣裙花了我三天时间。
5）她正在为玛丽设计漂亮的连衣裙。
6）她去年在设计大赛上获奖。
7）他们是工作努力的男装师。
8）今年我们引进了很多新设备。
9）在王师傅的帮助下，我设计了一件上衣。
10）借助于这些工具，我们修好了这台缝纫机。
11）他们一再强调质量的重要性。
12）这是男装的新流行趋势。

13）工人们正在操作缝纫机。
14）她很有设计天赋。
15）她们愉快地做着缝纫工作。
16）我们的服装有着精湛的工艺和优异的质量。（be of）
17）这两件衣服的区别是什么（在哪里）?
18）修理这台缝纫机花了我两个小时。
19）我们在安全方面做了很大改进。
20）这是今年流行的时装。

Conversation

>> We Are Tailors, Dressmakers, Cutters and Designers 我们是男装师、女装师、裁剪师和设计师

Conversation 1

A: I hear you graduate from a garment vocational school. Have you got a job?

B: Yes, I am a garment worker, now. I have got a job in a clothing company.

A: What do you think of the job?

B: The job is interesting. To tell the truth, that's just what I have long dreamed of job. I want to be a famous designer one day. I have to start at the bottom. I like the job very much. I am interested in anything concerning clothing and design. I love the company. I believe the knowledge and skill gained through studies and practice will be useful.

A: What do you think of the company?

B: The company is modern and beautiful. Our company has given us many opportunities for further study. We have the latest sewing machines and equipment. The technicians and engineers have invented many new machines. All the products they produced are very popular because of their modern design and excellent quality.

A: How did you do your work?

B: I work hard. I am a hard worker. I am good at my job. I always finish my work ahead of schedule. I do my best to improve the quality of my products. I believe quality is more important than quantity. But I want to change my life. I am learning design in a spare time school. My dream is becoming a designer in the future.

A: Are you a tailor or a dressmaker?

B: I am a dressmaker. I am working as a garment worker in the assembly line.

A: What does a dressmaker do?

B: A dressmaker is a worker who stitches clothes for women. I have learned a lot from the workers technicians, engineers and other staff members. I enjoy working every day.

Conversation 2

A: Mary has got a good job as a dressmaker in a clothing company.

B: I am pleased to hear the news. She has at last got a job. I hear her sister takes a part-time job in a factory. She works four days a week. Part-time work is generally hard to find.

A: What does she do?

B: She works as a designer. She has worked as a cutter for years in a clothing company. So she has rich experience in fashion design. She is very confident in her ability. Confidence in oneself is the first step on the road to success. She is confident to learn the subject well.

A: What does a designer do?

B: A designer is person who designs clothes. Her sister has a great talent for design. She has shown her great talent in this field. She draws, paints and makes patterns in her own way. She is different in style, color matching and material selection. She can design the most beautiful dresses. She won the first prize at the design contest last year. She was a diligent designer and deserved the prize.

A: It sounds great! We should learn from her.

New Words and Expressions

clothing company	服装公司	improve	改善
design contest	设计比赛	hard worker	勤奋的工人
generally	一般来说	excellent	出色的
assembly line	流水线	graduate	毕业
spare time	业余时间	diligent	勤奋的

deserve	值得的	work as	当，成为
sound	听起来	quantity	数量
learn from	向……学习	before schedule	提前
show one's talent	显示某人的才能	modern	现代的
hard to find	很难找	what do you think of	你认为……怎么样?
part-time	兼职的		

Reading Practice ❶

>> Mary's Family 玛丽的家

Mary is a dressmaker. A dressmaker is a worker who makes women's dresses. Her brother is a tailor. A tailor is a maker of garments, usually makes outer garments. Her father is a designer. Her sister is a designer, too. Her mother is a cadre and she is a manager of a clothing company. Mary and her brother work in a garment factory. They are all engaged in the garment industry. They all work hard. They are all skilled workers. They are better than the other workers at creative thinking. Their efficiency is much more than new workers. They have done great service to the garment industry. They love the industry and do their work with pleasure. They often help the other workers and teach them how to improve their skills. They sew and stitch many coats and dresses every day.

Every day after dinner, they talk about something interesting. They talk about fashion, clothing, the latest trend in clothes and leisure activities. They read books and magazines on fashion and clothes. They often read online journals and articles. Reading these books and magazines can extend their knowledge in different fields. Sometimes they exchange their own experience in their work. Her father has much experience in designing dresses. He is an experienced designer. He won many design prizes in the clothing contest last year. They show the same interest in fashion and leisure sports and talk about the topics with great interest. They often have the same opinions on the fashion. They find enjoyment in fashion. They are in the enjoyment of a happy life. What a happy family it is!

New Words and Expressions

maker	制造者
out garment	外衣
cadre	干部
manager	经理
skilled	熟练的
fashion	时装
have the same opinions	有同样看法
find enjoyment in	从……得到乐趣
show interest in	对……有兴趣
industry	工业，行业
enjoyment	乐趣
journal	刊物
do sth. with pleasure	做某事很高兴
leisure activities	休闲活动
exchange	交换
experience	经验
experienced	有经验的
show the same interest in	表现出同样的兴趣
be engaged in	从事
efficiency	效率
do great service to	对……帮助大
topic	话题
with great interest	怀着很大兴趣
be in the enjoyment of	享受着
extend	扩展

Reading Practice 2

A Fashion Designer 一位时装设计师

Emily is a designer. She works in a big company in the city. She has a good command of art and painting, the theory of clothes and English. She is good at stitching and sewing. She like reading all kind of books, magazines and journals on clothes and garments. She sometimes reads online books and journals. Reading these books, magazines and journals can extend her knowledge in different fields. She has a fashion concept of keeping the pace of the times. She graduated from a famous university five years ago. She studied in a famous college of fashion design and went to France to study fashion design for two years. She designs clothes with passion. It is the passion that motivates her to design all kinds of clothes. Her clothes are popular with the young people, especially young girls. She is going to hold personal exhibition recently. And she is considering going to America for further study.

Now she works together with a group of tailors, dressmakers, cutters and other workers. They are painting, making patterns, cutting out the clothes, sewing and pressing them well. They do their best to complete the work. The clothes designed by the team with experience and skills can for sure win the contest.They are determined to win the prize. They are sure they can design the best clothes and do a good job in the exhibition.

New Words and Expressions

have a good command of	通晓，精通	press	熨烫
theory	理论	team	团队
concept of	……的概念	for sure	肯定
keep the pace of the times	与时俱进	contest	比赛
passion	激情	make pattern	制板，出纸样
motivate	激励	be determined to	决心做
especially	特别是	consider	考虑
hold	举办	further study	深造

Notes to the Reading Practice

1. She has a fashion concept of keeping the pace of the times. 她有着与时俱进的时尚理念。
2. She designs clothes with passion. It is the passion that motivates her to design all kinds of clothes. 她用激情设计服装。正是这种激情激励着她设计各种服装。“It is...that...” 为强调句型。例如：It is the color of the garment that attracts my attention. 这件衣服的颜色吸引了我。
3. The clothes designed by the team with experience and skills can for sure win the contest. 这个既有经验又有技术的团队设计的服装肯定能赢得比赛。

Lesson Three

Text 1

>> Garment Collars 服装领子

A collar is a part of a shirt, a jacket or a coat that fits round your neck. It is a part of a garment at the neckline; usually it is sewed on as a separate piece. A collar is one of very important parts in a garment. Love beauty, everyone has it, the love of clothing is common to everyone. The quality of a garment mainly depends on it. It is pleasant to have a garment with a beautiful and fashionable collar for many young people. Sometimes it is the style of a collar that attracts people's attention. Some styles of collars impressed a lot of people. People can remember what the collars look like after a long time. The style of a collar keeps changing with the times.

Here are some collars. Some are round collars. Some are pointed collars. Some are stand-up collars and some are turned-down collars. This is a regular collar. That is a mandarin collar. Look here, you can find many wide-grooved tulip collars, wing collars, convertible collars, fur collars and flip collars.

If you want to make a garment fashionable, sometimes you should pay much attention to the collar of a garment. It is not easy to design a new and fashionable collar. The style of a garment shows the beauty of a garment. Different collars show different styles. Some collars of dresses should be carefully designed. The choice of collar reflects a person's taste in clothes. The choice of collar leaves a good or bad taste to people. But there is no accounting for taste. Sometimes the choice of collar is a subject of interest in some people. The new style of a collar holds a special attraction for young people. People are usually attracted by the style of a collar, especially young people and women. Maybe a style of collar is caught up and immediately become popular.

New Words and Expressions

collar	领子，衣领
fit round	适合地围着
neck	颈部
neckline	领围线
be fit round	适合地围着
be sewed on	缝在
separate	分开
part	部分
mainly	主要地
stand-up collar	竖领
turned-down collar	翻领
depend on	依靠
round collar	圆领
pointed collar	尖领
regular collar	普通领
no accounting for	无法解释的
leave a good taste to	给……留下好印象
accounting	解释
It is pleasant to do	做……是令人愉快的
fur collar	毛皮领
mandarin collar	中式领
wide-grooved tulip collar	厚罗纹领
wing collar	燕尾领
convertible collar	两用领
convertible	可转换的
flip collar	翻驳领
fashionable	时髦的
pay much attention to	非常注意
reflect	反映
choice	选择
taste	品位
beauty	美
a subject of interest	有趣的话题
be attracted by	被……吸引
attract	吸引
hold	具有
maybe	也许
a special attraction to	特殊吸引力
catch up	被人们接受
maybe	也许
immediately	立刻
common to	共有

Notes to the Text 1

1. A collar is a part of a shirt, dress or coat that fits round your neck. It is a part of a garment at the neckline; usually it is sewed on as a separate piece. 衣领通常是单

独缝制，装在衬衫、连衣裙、外套的领圈上合适地围绕在颈部的部分。

2. Some styles of collars impressed a lot of people. People can remember what the collars look like after a long time. 有些领子的款式给很多人留下深刻印象，人们过了很长时间还能记住它们的样子。

3. The style of a collar keeps changing with the times. 衣领的式样随时代变化而变化。

4. The choice of collar reflects a person's taste in clothes. 衣领的选择反映一个人的服装品位。

5. People are usually attracted by the style of a collar, especially young people and women. 人们经常被衣领式样所吸引，尤其是年轻人和妇女。

6. Some collars of dresses should be carefully designed. 有些连衣裙的衣领应该精心设计。

7. Love beauty, everyone has it, the love of clothing is common to everyone. 爱美之心人皆有之，对服装的爱好是每个人都有的。

8. But there is no accounting for taste. 人各有所好（嗜好是无法解释的）。

9. The new style of a collar holds a special attraction for young people. 新式样衣领对年轻人有着特殊吸引力。

10. Maybe a style of collar is caught up and immediately become popular. 也许一种衣领款式被人们接受并很快流行起来。

Text 2

>> A Garment Collar 一款服装衣领

Today we are going to learn a collar of a garment. Look, this is a collar. It is a pointed collar. The collar is the most popular style among the young people. It is beautiful and fashionable. Beauty has an attraction for all. It is the collar that attracts many girls and young ladies.That is a regular collar. It is out of date in the market.

But it is believed that it will be back in fashion. That's the trend. It is going to happen again in a few years.

The style of the garment is still popular in the market except for the collar. The collar is out of date now. If we change the style of the collar. It is still a fashionable garment. It will enjoy great popularity. We can see from the above example that a collar is a practical and nice part of a garment. It is a small part but it is very important to a garment. It plays an important role in a garment. Sometimes the overall advantage of a coat lies in the collar.

The usage and the function of a collar are not the same. Sometimes people pay attention to its function but sometimes people neglect it. The style and shape are changed with the time. Collars attract people's attention. Many people are interested in the style of a garment, but some people are not interested here. The collar of a garment shows the important and practical part of a garment. Sometimes the quality of a garment mainly depends on the style, color, the way of stitching it. Some people like a garment at the first sight of the collar. Different collars are suitable for different people. So designers pay much attention to collars to meet the tastes of different people. Designers are offering beautiful style and better quality collars and garments. They want all the designs to be very attractive and fashionable. They want to make people's lives better.

New Words and Expressions

pointed collar	尖领
out of date	过时
it is believed that	人们相信
be back in fashion	重新流行
that's the trend	这就是流行趋势
except for	除了
enjoy great popularity	很受欢迎
attraction	吸引力
overall advantage	整体优势
lie in	在于，体现在
lady	女士
neglect	忽视
practical	实用的
above example	上述例子
play an important role in	在……起重要作用
the usage and the function	用途和功能
pay attention to	注意
attract one's attention	吸引某人的注意
mainly depend on	主要靠
the way of doing	做某事的方法
at the first sight of	一看见
be suitable for	对……适合
meet the tastes of	适合……的品位
offer	提供

Notes to the Text 2

1. Beauty has an attraction for all. 美对所有人都有吸引力。

2. It is out of date in the market. 这种衣领在市场上已经不流行了。

3. The style of the garment is still popular in the market except for the collar. 除衣领过时外，这款服装在市场上仍然流行。except for+名词（短语），except that+句子。例如：This suit fits me well except that the trousers are too long. 除了裤子太长外，这套服装很适合我。He answered all the questions except for the last one. 除最后一题外，他回答了所有的问题。

4. ...are not the same. 不一样。注意same永远和定冠词“the”连用。

5. Designers are offering beautiful style and better quality collars and garments. 设计师们正在提供款式漂亮、质量优异的服装和衣领。

6. All their designs are very attractive. 他们所有的设计都非常吸引人。

>> Substitution Drills 替换练习 <<

<table>
<tr><td rowspan="3">1. The style of</td><td>a collar</td><td rowspan="3">keeps changing with the times.</td></tr>
<tr><td>a jacket</td></tr>
<tr><td>a shirt</td></tr>
</table>

<table>
<tr><td rowspan="3">2. Here are collars. Some are</td><td>round</td><td rowspan="3">collars.</td></tr>
<tr><td>pointed</td></tr>
<tr><td>regular</td></tr>
</table>

<table>
<tr><td rowspan="3">3. The style of</td><td>a collar</td><td rowspan="3">keeps changing with the times.</td></tr>
<tr><td>dresses</td></tr>
<tr><td>coats</td></tr>
</table>

4. If you want to make a garment fashionable, sometimes you should pay attention

<table>
<tr><td rowspan="3">to</td><td>the collar of a garment</td><td rowspan="3">.</td></tr>
<tr><td>the collar's color</td></tr>
<tr><td>the collar's material</td></tr>
</table>

5. The style / The color / The appearance of a garment shows the beauty of a garment.

Exercises to the Texts 课文练习

1. 抄写熟记本课单词和词组
2. 翻译下列词组

1）round collar
2）pointed collar
3）depend on
4）be attracted by
5）the choice of collar
6）pay attention to
7）the style of a collar
8）the beauty of a garment
9）fit round
10）a subject of interest
11）flip collar
12）wing collar
13）catch up
14）fur collar
15）out of date
16）except for
17）above example
18）attract one's attention
19）be suitable for
20）at the first sight of

3. 将下列短语翻译成英语

1）两用领
2）中式领
3）大翻领
4）尖领
5）领围线
6）衣领的质量
7）衣领的式样
8）使服装更时髦
9）市场上
10）衣服的质量
11）起重要作用
12）衣领的选择
13）具有特殊吸引力
14）翻领
15）给……留下好的印象
16）漂亮的款式
17）忽视
18）成为流行
19）改变款式
20）精心设计的

4. 将下列句子译成英语

1）请将衣领缝在领圈上。
2）服装的质量是很重要的。
3）服装的款式随着时间变化而变化。
4）服装的选择反映人的品位。

5）人们常常被服装的款式吸引。
6）服装的选择是个有趣的话题。
7）让我们学着使我们的服装更时髦。
8）这款式表现了服装的美。
9）人各有所好。
10）女孩子经常被时装吸引。
11）新款服装对年轻人有特殊的吸引力。
12）美对我们大家都有吸引力。
13）不同的服装适合不同的人。
14）这些服装都是精心设计的。
15）我们应该非常注意质量。
16）不要忽视安全。
17）那款裙子过时了。
18）我们的服装在市场上很受欢迎。
19）这就是缝纫衣领的方法。
20）我一看见这件夹克就喜欢它。

Conversation

>> Dialogues About Clothes 有关服装的对话

Conversation 1

A: What do you think of the new jacket?

B: It looks nice, but the jacket is too thin. it's not as heavy as the black one. You know, one of the functions of a garment is cold resistant, it should be heavier. I like to wear a thicker jacket. A thicker jacket keeps you warm.

A: Yes, I think it looks good on you. It fits you nicely.

B: Really! It looks good on me? I don't think so. How do you think of that jacket? I bought one last week. This jacket looks thicker than those ones. And it appears more younger.

A: I think it's out of style. It's not the fashion of this year. The fashion does not take.

B: You are right. I find the collar is a little tight. I'd like to exchange it for another one. Could you please go with me?

A: OK, Let's go.

Conversation 2

A: How are you doing with your design in the vocational school?

B: Very well. Our teacher Mr. Wang is one of the experienced designers in our school. He is an excellent designer. Every year he designs all kinds of dresses for our models. Last week he taught us how to design all kinds of collars, especially the fashion for women's wear. Fashions for men's clothes change less frequently than fashions for women's clothes.

A: You are right. Students should be encouraged to make full use of time to learn more and find beauty of nature. Does he draw as well as make patterns?

B: Yes. He draws watercolors and make perfect patterns. And sometimes he shows us how to design, draw, and make patterns. Everybody thinks he is a good teacher and designer. He can fashion the dress to your figure.

New Words and Expressions

what do you think of...	你觉得……怎么样?
thin	瘦的
heavy	厚的
cold resistant	御寒
thicker	厚一点的
keep sb. warm	使某人保暖
look good on sb.	某人穿好看
tight	紧
nicely	非常
how do you think of...	你认为……怎么样
out of style	过时
be not the fashion	不流行
exchange	换
does not take	不受欢迎
model	模特
how are you doing with your design in...	你设计学得怎么样?
less	较少，不如
frequently	频繁
encourage	鼓励
make full use of time	充分利用时间
nature	自然
vocational school	职业学校
experienced	有经验的
be very experienced in	在……很有经验
as well as	以及
make pattern	出样板
show sb. how to do	教某人怎么做
fashion	使适合
fashion ...to one's figure	把衣服做得合某人的形体

Reading Practice 1

Different Collars 不同的衣领

We have many garments here. Every garment has a collar. A collar is a part of a garment. It is the "eye" of a garment. The collars are different. The styles of the collars are different, too. Some are regular collars, some are round collars, some are pointed collars, and some are high collars. Some are the spring fashions and some are the autumn fashions. The design and decoration of the collars show their great improvement.

Young people always pay attention to the collars of garments. They focus on the collars of garments. Every year all the companies and firms do their best to design many new collars to meet the customers' needs. They offer the latest designs to customers. Different collars have different styles. The style of a collar shows the style of a garment. Different people like different collars. Beautiful and fashionable collars attract people to buy. Sometimes collars are the attraction to some people. Young customers generally take great interest in collars. Some garment stores and supermarkets always have many kinds of collars to choose from. Everyone has his own choice. Every collar has its own taste. This collar is not in your taste, but the other collar may be in your taste. Many people pay attention to the style of collars, especially girls, young people and women. The style of the collar reflects a person's taste in clothes. So the choice of collar is very important.

New Words and Expressions

do one's best to do	尽某人最大努力	attract	吸引
take great interest in	对……感兴趣	firm	公司，商号
spring fashion	春季流行款	decoration	装饰
choose	选择	focus	注意力放在
reflect	反映	choose from	供选择
attraction	吸引人的地方	focus on	目光集中在

Reading Practice 2

Collars 衣领

We design, make and stitch many garments every day. We can see many beautiful garments in the department store, supermarkets, clothing store and Brand clothing store. Generally speaking, every garment has its own collar. Different people like different collars. To some extent, the taste for clothes is closely related to one's cultural background. The styles of garments are the main points. People sometimes pay much attention to them. The collar may not be in your taste, but it may be the others' favor. Many clothing stores and department stores have many different styles of collar in the same garments. People have many choices in the collar. The styles of garments are changed with the times. Some collars are fashionable and popular nowadays, but maybe they are out of date next year. So their popular life is short. A few years later, they are received into people's favor.

There is a new fashionable collar in the market recently. People think the style of collar is estimated to last four years. The collar of a garment is one of the most popular topics with young people, especially girls, young ladies and women. We must learn how to design and make all kinds of collars to meet the needs of different people. And we should do our best to enrich our garments and clothes.

New Words and Expressions

extent	程度
to some extent	某种程度上
estimate	估计
cultural	文化的
background	背景
closely	紧密地
relate	和……关联
brand	牌子
Brand clothing store	品牌店
generally speaking	一般来说
favor	喜爱，偏爱
nowadays	现在
popular life	流行期
later	以后
topic	话题
be received into one's favor	受某人的青睐
last	延续
enrich	丰富

Lesson Four

Text 1

Introducing a Garment Factory 介绍一家服装厂

This is our factory. It is a clothing factory. It is not very large, but it is modern, clean and tidy. We started the factory forty years ago. The factory produces large quantities of high-quality garments every year. Our factory is an enterprise specializing in the import and export of clothes. We trade with nearly all the countries and regions in the world. We receive thousands of businessmen from all over the world every year. They like our products very much. Our products sell well all over the world.

There are many workshops in our factory, including cutting workshops, sewing workshops, pressing and ironing workshops and packing workshops. Besides these workshops, we have warehouses, design rooms, showrooms, office buildings and other buildings. The other two new workshops are still under way and perhaps they will be completed in a month or two. But the complete equipment of the new workshops will take another month.

We have a lot of modern equipment in our factory. We have many sewing machines, steam irons and other kinds of new instruments and machines in our factory. Some of them are made in our country and some of them are introduced from other counties. We have invented many machines and equipment. They play a very important role in our job. The modern equipment in our factory works perfectly. Traditional techniques have gone through many changes in the past few years. We have developed techniques which are superior to those used in most factories.

Every year we invite businessmen, designers, engineers and technicians of other countries and regions to visit our factory. We exchange idea of requirements, views and experience with them. We get information from Internet. We enter various competitions every year. We have won many prizes in the past few years.

New Words and Expressions

tidy	整洁的	equipment	设备
enterprise	企业	steam iron	蒸汽熨斗
specializing in	专门做	instrument	器械
import and export	进出口	play a very important role in	起重要作用
trade	贸易	modern	现代的
nearly	几乎	perfectly	完美地
region	地区	technique	技术
including	包括	go through	经历
press	压烫	develop	开发
iron	熨烫	technique	技术
pack	包装	superior to	优于
besides	除……之外	engineer	工程师
warehouse	仓库	exchange	交换
design room	设计室	technician	技术员
showroom	陈列室	idea of requirement	需求量概况
be still under way	还在建设中	view	意见
perhaps	也许	experience	经验
complete	全部的	get information	获取信息
enter competitions	参加比赛	Internet	互联网
businessman	商人	office building	办公楼
enter	参加	take another month	还需一个月
start	创办	produce	生产
large quantities of	大量的	high-quality	优质
sell well	畅销	receive	接待

Notes to the Text 1

1. The other two new workshops are still under way and perhaps they will be completed in a month or two. 另外两个新车间还在建设中，可能一两个月后建成。

2. But the complete equipment of the new workshops will take another month. 但是新车间全部完成设备安装还需要一个月。

3. Traditional techniques have gone through many changes in the past few years. 过去几年传统技术经历了很多变化。

4. We have developed techniques which are superior to those used in most factories. 我们开发了许多优于其他厂的技术。

5. ...exchange views and experience with... 与……交换意见和经验

6. get information from Internet 从网上获得信息

Text 2

>> A Garment Factory 一家服装厂

Today I am going to visit a garment factory. It is a modern factory. It is clean and tidy. The factory was built 40 years ago. Now it looks like a beautiful park. The workers have made outstanding achievement with great courage, confidence and creativity in the past 40 years.

The factory is an enterprise specializing in the import and export of clothes. They have many skilled workers, cutters, tailors and dressmakers. They have a lot of good designers in their factory. Their workers produce at a very fast rate. Their high rate comes from their hard work. Their workers learn and master traditional and modern techniques. They have developed and improved techniques which are most suitable for their products. They have many workshops in their factory, including cutting workshops, sewing workshops, pressing and ironing workshop and packing workshops. Besides these workshops, they have warehouses, design rooms, showrooms, office buildings and other buildings. They design and make all kinds of men's wear and women's dresses. But they also make all kinds of children's wear, sports wear and underwear. Their products are beautiful and fashionable and become popular among young people. Some of them are very popular with foreign friends. Many of their products come into wear, that is to say their garments often become a fashion guide. Their products are elegant in taste dress, and young ladies and women admired the elegance of the lady's clothes. Whenever their new products are on the market many people will go to buy them. With the development of online sales, they opened many shops on the Internet and their products are for sale online and have been widely welcomed. I am sure I can get a lot of benefit from the visit.

New Words and Expressions

look like	看起来像
outstanding achievement	了不起的成就
courage	勇气
confidence	信心
creativity	创造力
specialize in	专门
import	进口
export	出口
at a very fast rate	以非常快的速度
rate	速度
master	掌握
be suitable for	适合
besides	除……之外
elegant	优雅的
elegance	优雅
admire	欣赏
improve	改善
be suitable for	适合
men's wear	男装
women' dress	女装
children's wear	童装
underwear	内衣
come into wear	成为流行式样
that is to say	即，也就是说
guide	导向
whenever	无论何时
fashion guide	时尚引导
with the development of	随着……的发展
online sales	网上销售
get benefit from	从……获益

Notes to the Text 2

1. The workers have made outstanding achievement with great courage, confidence and creativity in the past 40 years. 在过去的40年里，工人们以巨大的勇气、信心和创造力取得了了不起的成就。

2. The factory is an enterprise specializing in the import and export of clothes. 这家工

厂专门从事进出口服装。

3. Their workers produce at a very fast rate. Their high rate comes from their hard work. 工人们的劳动生产率很高。他们的高效率来自他们的辛勤劳动。

4. They have developed and improved techniques which are most suitable for their products. 他们开发和改进了很多适合他们产品的技术。

5. Many of their products come into wear, that is to say their garments often become a fashion guide. 他们的许多产品成为流行式样，也就是说他们的服装经常成为时尚导向。

6. Their products are elegant in taste dress, and young ladies and women admired the elegance of the lady's clothes. 他们的女装品位高雅，年轻女士和妇女们都很欣赏他们服装的雅致。

7. With the development of online sales, they opened many shops on the Internet and their products are for sale online and have been widely welcomed. 随着网上销售的发展，他们在互联网上开了很多店并受到了广泛欢迎。

8. I am sure I can get a lot of benefit from the visit. 我相信我一定能从这次参观中获益很多。

>> Substitution Drills 替换练习 <<

1. We specialize in [clothes / sewing machines / caps and hats / men's wear].

2. [The other two new workshops / The new schools / The buildings / Discussions] are still under way.

3. Our factory is an enterprise specializing in the import and export of [clothes / dresses / jackets / suits].

4. We exchange [views / experience / equipment / ideas] with foreign friends.

5. We have developed [techniques / sewing machines / products] which are superior to other factories.

6. [The equipment plays / The computers play / The new technique plays] a very important role in our job.

Exercises to the Texts 课文练习

1. 抄写并熟记本课短语和词组
2. 翻译下列短语

1）superior
2）go through
3）import and export
4）specialize in
5）be suitable for
6）underwear
7）that is to say
8）online sales
9）high rate
10）modern technique
11）exchange views
12）get information from Internet
13）various competitions
14）traditional techniques
15）steam iron
16）enter competition
17）work perfectly
18）be still under way
19）get benefit from
20）fashion guide

3. 将下列词组译成英语

1）干净整洁
2）交流经验
3）起重要作用
4）办公楼
5）童装
6）现代化设备
7）成为流行设计款式
8）交换意见
9）其重要作用
10）传统技术
11）先进技术
12）开发技术
13）在过去的几年里
14）参加比赛
15）即
16）外国商人
17）经历
18）获得信息
19）开办网上商店
20）男装

4. 翻译

1）我们厂专门从事进出口服装。
2）我们的新整烫车间正在建设中。
3）他们引进了新式缝纫机。
4）传统技术很重要。
5）我们每年邀请外国设计师来访问。
6）我们从互联网上获取最新的时装信息。
7）互联网在我们工作中起了很大的作用。
8）Tom 开发了一种新压烫机。
9）这家服装厂看起来像个公园。
10）另一家厂还在建设中。
11）我们开发了许多新技术。
12）我们与外商交换了意见。
13）工人们的高效来自于辛勤劳动。
14）他们为孩子们设计了很多新款服装。
15）我们的许多产品成为流行式样。
16）我相信我们会从这次访问中获益很多。
17）我们邀请外商参观我们公司。
18）传统技术在过去几年里经历了许多变化。
19）除了英语我还学了设计。
20）我们有很多有技术的工人。

Conversation

>> Visiting a Sewing Workshop 参观缝纫车间

Conversation 1

A: This way, please. This is our workshop. We have five assembly lines in it. These assembly lines are modern and efficient. We built the assembly lines two years ago. They meet the needs of production. Here we have the best equipment and machines in the world. There are more than 100 workers working here. They are all skilled workers with rich production experience.

B: How large it is! You have so many new sewing machines, steam irons, computers and other equipment.

A: Yes. We have sewing machines and steam irons controlled by computers. They are all developed by the engineers of our factories. We also have many sewing machines introduced from foreign countries. They work perfectly and play a very important role in our jobs.

B: What about your workers? Are they qualified for their positions?

A: Yes. These workers are of high quality. They are specially trained for the kind of work. Every worker is well qualified for his or her position. They do their best to work.

B: You have so many good workers and new machines. I am sure you can make high quality clothes. I'm really sure about it.

A: Thank you. I am confident that our workers will produce more fashionable clothes.

Conversation 2

A: Would you please show me how to stitch the pocket? I'd like to learn it. Is it easy?

B: All right. I'll show you what to do. But it's not easy to do this job without experience. First, fold down the top edge of the material of the pocket, and then fold down both sides of the pocket and the both corners of the pocket. We press the pocket with steam iron.

A: Now we have the pocket formed, all right?

B: Yes. We put the pocket on the front piece and stitch along the edge of it. At the beginning and the end, we shall backstitch it to reinforce the pocket.

A: Oh, the pocket fits well. It's so nice. I can do it myself.

B: Do you follow what I said and what I did? You have to learn how to do it, and you need more practical training. You should be received strict vocational training if you want to be a skilled worker.

A: Yes, I do. Thank you for your help.

B: You're welcome.

New Words and Expressions

assembly line	流水线	have...formed	使……成型
perfectly	完美地	stitch along the edge	沿着……的边缘缝
be qualified for	胜任	backstitch	来回的缝
position	岗位	reinforce	加固
fold down	折倒	be of+adj.	具有……的性质
production experience	生产经验	practical	实际的
training	训练	strict	严格的
fit	适合	vocational training	职业培训

Reading Practice 1

Visiting a Garment Factory
参观一家服装厂

This is our factory. It was built according to the design of famous engineer. It was built many years ago. It is large, clean and modern. It is not a textile factory. It is a garment factory. There are many workshops, office buildings and other houses in our factory, they are cutting workshops, sewing workshops and packing workshop. We equip our workshops with new sewing machines, steam irons, cutters and other equipment. We have also invented equipment that other factories don't have. They are all controlled by computers. The equipment can greatly improve the quality and production efficiency. With these equipment, the output has increased many times.

Workers must receive strict training before they go on duty. The workers are trained to learn professional skills. There are many kinds of skills that have to be learned before one can do this job. We have a regular job training for our workers every year. We have modern ways to teach the workers how to learn the skills. They can also learn from their masters, teachers and learn from each other. They can not only learn traditional techniques but also learn modern techniques. The workers can get information from the Internet and ask for help on the Internet when they have difficulties. They work hard and do well on the work. They are confident that they will make the best products for customers. Their promise to customers is that their products will be what they expect. Both the quality and style can meet the needs of customers.

New Words and Expressions

be built according to the design of sb.	根据……的设计建造
textile	纺织
equip...with	给……配备
control	控制
output	产量
efficiency	效率
customer	顾客
regular job training	定期职业培训
promise	承诺
expect	期待的
with the equipment	有了这些设备
go on duty	上岗

Reading Practice 2

Our Factory 我们厂

Welcome to our factory. Now, let me show you around our factory. This is our garment factory. It is a new and modern factory. It was built according to the design of a famous designer. It was built on a garden model, so the factory looks like a beautiful garden. We have many workshops, office buildings and other buildings in our factory. Our factory covers an area of 1000 mu. The area of our garden is about 10000 square meters. Our factory is not only beautiful, but also modern. We have the most modern machines and equipment. Most of them are invented, designed and made by ourselves.

There are several sewing workshops, cutting workshops, and packing workshops in our factory. We have a teaching building, it is our teaching base. In our teaching building, we equip with modern sewing machines, cutters, steam irons, and other equipment. There is also multimedia teaching equipment in the teaching building. Every year we train hundreds of skilled workers and designers in our teaching base. They can learn not only traditional techniques but also modern techniques. They can read and get useful information on the Internet. They can communicate with foreign workers and friends all over the world on the Internet. The workers in our factory work hard and complete the task every month. They produce the most fashionable garments to meet the customers' needs.

New Words and Expressions

on a garden model 按花园模式
cover an area of 占地……
teaching base 教育基地
equip with 配备有
multimedia 多媒体
traditional technique 传统技能
communicate with 与……交流
be built according to a design of a famous designer 由一位著名的设计师设计建造
be built according to a garden model 它采用花园模式建造
specializing in 专门做
be still under way 还在建设中
enter competitions 参加比赛
go through many changes 经历了许多变化
superior to 优于
at a very fast rate 以非常快的速度
come into wear 成为流行式样
fashion guide 时尚引导
with the development of 随着……的发展
fold down 折倒

Lesson *Five*

Text 1

>> The Garment Vocational School 服装职业学校

This is our school. The name of our school is “The Garment Vocational School” . Our school is mainly training the students for designing and making up all kinds of clothes. We train garment workers and technical cadres regularly in short courses so that they continue to enrich their knowledge and new skills.

Our school is not a large one. There are several departments in our school. There are about eight hundred students altogether including teachers, designers, staff members, coaches and students. This is the only one kind of school in our city, so it is very important and useful. The school is popular with the parents and students.

There are twenty classes in our school. We have lessons on art, design, computer program, clothing material, and specialized knowledge, we also have Chinese, mathematics, English and CAD. Usually we have lessons in the morning and do some cutting, stitching and pressing in the afternoon. Every student is equipped with a computer and pad. The students often get information on the Internet. Sometimes it benefits us greatly. Our school places a high value on personality cultivation besides on education, so everyone in our school can choose what they are interested to learn. We believe the skills gained through studies in our school will certainly be in high demand in the future.

The students in our school study very hard. They keep all that the teachers taught in their minds. They realize that if they really want to succeed, they will have to work hard. Our designs are exhibited in the show or fair every year. We also design dresses and fashions for garment shops and other firms. After graduation the students are going to garment factories，clothing companies and some of them are likely to enter different supermarkets, clothing stores. Some of them may go to universities or go abroad for their further studies. Studying abroad will make the students learn a lot from foreign countries and also make them much more independent.

New Words and Expressions

technical	技术的
train	培养，培训
technical cadre	技术干部
regularly	定期
short course	短训班
continue	继续
enrich	扩大
department	部门
altogether	一共
staff member	职员
be popular with	受……欢迎
computer program	计算机编程
clothing material	服装材料
specialized	专业的
certainly	一定
further study	深造
university	大学
knowledge	知识
benefit sb. greatly	对某人很有利
be likely to do	想做
firm	商号，公司
graduation	毕业
place a high value on	很注重
personality cultivation	个性培养
gain	获得
be in high demand	有很大的需求
several	几个
coach	实习老师
equip with	配备
pad	平板电脑
benefit	有利
education	教育
keep in mind	牢记
go abroad	出国

Notes to the Text 1

1. all that the teachers taught 老师所教的

2. Our designs were exhibited in the show or fair every year. 我们的设计每年都会在展览会或交易会上展出。

3. Some of them may go to universities or go abroad for their further studies. 有的学生会上大学或出国深造。

4. Our school place a high value on personality cultivation besides on education. 我们学校除了注重教育外也重视个性培养。

Text 2

>> A Garment Vocational School
一所服装职业学校

This is our school. It is a garment vocational school. The education goal of our school is to train clothing talents for our country and cultivate management talents for our country. The teachers in the school do their best to develop the students' talents. The students in the school learn painting, sewing, stitching, cutting out clothes, embroidering, pressing and ironing. They also learn how to design and make up all kinds of clothes. They learn all subjects including Chinese, mathematics, English, art design, computer program, clothing materials and specialized knowledge. They also learn other foreign languages such as French, Japanese and Russian. They learn how to design clothes and make paper patterns.

The teachers of our school think of their students as their friends, so they are popular with the students. The teachers encourage the students to innovate and develop their creative ideas. We put emphasis again and again on the importance of traditional skills and creative ideas, neither should be neglected. Our school pay much attention to the quality education. Our school places a high value on personality cultivation besides on education. The students are interested in creative idea. We encourage the students to make full use of computer and other modern technology. Wherever you go in the school you can see the students are talking about all kinds of new ideas and communicating with each other in English. Every student in the school can choose what they are interested to learn. Every year hundreds of students graduate from the vocational school. They are all skilled and qualified for the positions they will get. Our school has gone through many changes in recent years, and has become a famous garment vocational school.

New Words and Expressions

vocational	职业的
educational goal	教育目标
train talents for	为……培养人才
cultivate	培养
develop one's talents	开发某人的才能
make up	制作
embroider	刺绣
specialized	专业的
Russia	俄语
quality education	素质教育
creative idea	创新思维
innovate	创新
encourage	鼓励
make full use of	充分利用
place (lay, put) emphasis on	强调
emphasis	强调
again and again	再三
neglect	忽视
communicate with	与……交流
wherever	无论哪里
place a high value on	注重
personality	个性
graduate from	毕业于
be skilled for	熟练于
be qualified for	胜任
go through	经历

Notes to the Text 2

1. ...cultivate management talents for... 为……培养管理人才

2. ...what they are interested to learn学习他们所感兴趣的东西。what 关系代词 + 句子=all that 表示"所……"的意思。例如：Is this what you need? Maybe what he needed was a coat. Will you show me what you bought?

3. The teachers of our school think of their students as their friends, so they are popular with the students. 我们学校的老师们把学生当作自己的朋友，因此深受学生们的喜爱。

4. The teachers encourage the students to innovate and develop their creative ideas. 老师们鼓励学生们创新并开发他们的创新思维。

5. We put emphasis again and again on the importance of traditional skills and creative ideas,either should be neglected. 我们再三强调传统技术和创新思维的重要性，两者缺一不可。

6. Our school pay much attention to the quality education. 我们学校非常重视素质教育。

7. They are all skilled and qualified for the positions they will get. 对于他们将得到的岗位，他们都是熟练且合格的。

8. Our school has gone through many changes in recent years, and has become a famous garment vocational school. 最近几年我们学校经历了许多变化，成了一所著名的服装职业学校。

Substitution Drills 替换练习

1. The school / factory / shop is mainly for designing and making all kinds of clothes.

2. They design dresses / fashions / jackets for shops and supermarkets.

3. Some of them may go to universities / go to college / go abroad for further study.

4. I'd like to try on this dress / shirt / jacket / coat.

5. They keep all that the teachers / workers / coaches taught in their minds.

6. Our designs / dresses / clothes are exhibited in the show or fair every year.

Exercises to the Texts 课文练习

1. 抄写并熟记本课短语和词组

2. 翻译下列短语

1）all kinds of
2）be popular with
3）go abroad
4）all that
5）make up
6）enrich their skills
7）the only one
8）on the Internet
9）in high demand
10）in the future
11）staff member
12）the only one kind of
13）keep...in one's mind
14）have lessons on
15）benefit sb. greatly
16）in the show
17）quality education
18）creative idea
19）encourage sb. to do sth.
20）put emphasis on

3. 将下列短语译成英语

1）服装厂
2）技术学校
3）师生员工
4）受欢迎
5）配备
6）进一步深造
7）非常注重
8）创新思维
9）鼓励学生创新
10）毕业后
11）专业知识
12）出国
13）服装材料
14）服装公司
15）开发某人的才能
16）经过学习获得技能
17）个性培养
18）把……当作
19）胜任
20）熟练

4. 翻译

1）我们的时装在交易会上展出。
2）玛丽去年出国留学了。
3）我们班一共有30位同学。
4）计算机辅助设计对我们很重要。
5）Tom牢记老师所教的。
6）服装材料是一门很有用的课程。
7）我们学校设计了很多服装。
8）我们的服装很受顾客的欢迎。
9）他们公司有很多部门。
10）他们的产品很受年轻人的欢迎。
11）我们可能会出国深造。
12）他们都是熟练工人。
13）你能把你设计的服装给我看一看吗?
14）我们服装厂最近几年经历了很多的变化。

15）老师们鼓励学生们努力学习服装设计。

16）他们为这家超市设计了很多款夹克。

17）我想试试这件上衣。

18）他们把实习老师的话牢记在心。

19）他们非常注重素质教育。

20）学生们对创造性思维非常感兴趣。

Conversation

>> Fine Quality 优质

Conversation 1

A: This is our showroom. We display many kinds of garments, including coats, jackets, shirts, blouses, dresses and so on.

B: I have heard about the garments displayed here and I want to see the quality.

A: We take pride in our quality. All the garments here are different from what you saw last time. Look at this one. The material is good. The design is fashionable, the style is the fashion of the day and the stitching is excellent. You know. Quality is something we never neglect. Quality is more important than quantity. We aim at quality rather than quantity. So everybody in our factory pays attention to quality of products.

B: We are deeply impressed by the quality of your garment. Thank you for your introduction.

A: You are welcome.

Conversation 2

A: Let me show you around our workshop. Here we display a great variety of clothes.

B: We have no idea whether the quality is good or not. Could you please tell me something about it?

A: Our products are well known for their fine quality. Look at the business suit, please. The suit is of an excellent design. It has a very good cut. The stitch is superb. We are particular about shaping and pressing. So the quality is popular with people at home and abroad.

B: I'm wondering whether the cloth will shrink after wash.

A: No. We guarantee the quality of the material. Remember to dry clean it only.
B: I believe the quality of the suits will leave a deep impression on customers.
A: Thank you.

New Words and Expressions

hear about	听说	have a very good cut	裁剪得体
quality is king	质量是第一位的	superb	上乘的
take pride in	对……感到自豪	be particular about	特别注意
stitching	缝纫	shaping	造型
aim at	目标是	shrink	缩水
rather than	而不是	wonder	想知道
have no idea	不知道	guarantee	保证
be popular with	受欢迎	leave a deep impression on	留下深刻印象
be deeply impressed	对……留下深刻印象	promise	许诺
a great variety	大量的	top quality	优质
be of	具有……品质		

Reading Practice ❶

Our Garment Technical School
我们服装技术学校

This is our school. Our school is a garment technical school. It is a vocational school. Our teachers teach us how to design and make all kinds of garments. We study cultural knowledge, computer program traditional and modern skill for designing, cutting and sewing in our school. In addition, we also learn many foreign languages. We learn about information technology.

We learn many foreign languages such as English, Japanese, French, and Spanish and so on. Our school equips us computers and audio-visual aids for language teaching in every classroom. We can get a lot of information we need on the Internet. Sometimes we communicate with the students in other schools on the Internet.

The teachers in our school are all skilled and experienced in designing. Many teachers won many prizes in the contest. They teach us how to draw pictures and make paper patterns; they teach us how to choose garment materials. The teachers of our school communicate their ideas about clothes and design very clearly to the students. They have strong sense of responsibility in work. They teach us design all kinds of clothes to meet the needs of different people. We learn a lot from them. The school call on us to innovate and develop our creative ideas. We benefit a lot from the experience of innovation. They keep on practice every day. We know that we can only acquire skill in something by constant practice. Now many of us can cut out, stitch, and press all kinds of clothes skillfully on our own. We finished all the task which every one thought it impossible. We love our school, and we love our teachers.

New Words and Expressions

vocational school　职业学校
cultural knowledge　文化知识
information technology　信息技术
program　程序，方案
Spanish　西班牙语
audio-visual aids　语言教学视听器材
communicate sth. to sb.　传递给
clearly　清楚地
sense　感觉
responsibility　责任
strong　强烈的
creative idea　创新意识
on one's own　独自地
communicate...to　传达
communicate...with　与……交流
make paper patterns　出纸样
sense of responsibility　责任感
call on　号召
task　任务
on one's own　独立，靠自己
impossible　不可能的
experienced　有经验
innovate　创新，引入
innovation　创新
skillfully　熟练地
in addition　除此之外

Reading Practice 2

Our Garment Vocational School
我们服装职业学校

Here is our school. It is a garment vocational school. Our school is to train garment professionals for our country. The teachers of our school are all skilled and experienced, some of them are of great learning. They teach our students carefully. The students learn all useful subjects and traditional and modern skills for sewing, stitching and designing. They master all kinds of foreign languages and foreign trade knowledge. In our school, what impressed our students most are various activities, where we can learn a lot of useful and practical skills.

Now the students have free access to computers in classroom, which brings much convenience to teaching and learning. They can communicate with foreign businessmen and friends on the Internet. They can exchange their ideas and experience on clothes and design with their classmates and friends abroad. We are very grateful to our school and our teachers for providing us with such a good place to learn so many subjects. Every year our school train hundreds of tailors, dressmakers, and designers, some of them become foreign trade staff and businessmen. After graduation, the students go to garment factories, clothing companies and supermarkets. Emphasis on education and the rapid growth in the needs of fashionable garments make the market better and better. Some of them go abroad for further study. They love our school and our teachers.

>> New Words and Expressions <<

professional 专业人员
be of great 具有……品质
master 掌握
communicate with 与……交流
businessman 商人
impress 表达
various 各种各样的
activity 活动
rapid 迅速
growth 成长

emphasis 强调
emphasis on education 重视教育
make...better and better 使……越来越好
staff 职员
graduation 毕业
free access to 可以使用
convenience 方便
teaching and learning 教和学
exchange idea and experience 交换想法和经验

Lesson Six

Text ❶

>> Measurements Taking 量体

A: Good afternoon, Madame. What can I do for you?

B: I'd like to buy a blouse. I want to buy something good. Could you please show me some blouses to choose from?

A: Today blouses come in many colors, styles and prices. Which style do you prefer, traditional or modern? How do you like this pattern? It's the latest style.

B: Oh, This blouse is quite all right except for its color. I don't care for the color. The tastes to colors vary from person to person. What is suitable for others may not be so for me. Can you please show me some other styles? You know, we should spare no effort to beautify our clothes.

A: Certainly, how about this one? It is designed by our famous designer. The breast of the blouse has a beautiful design of flowers. Many young women find the design attractive. The blouse owes much to the decoration. The main attraction of the blouse is here.

B: All right. May I try it on?

A: Of course. The fitting-room is over there. This way, please.

B: I like the style but it's a bit larger for me. I'd like to order a blouse.

A: OK. Let me see. It's too loose round the waist. The sleeves are a little too long. I'll take your measurements. I'll have it styled by a famous designer.

B: Thank you.

A: Now let's start taking measurements. Coat length 60 cm, chest 100 cm, shoulder 42 cm, sleeves 58 cm, collar 39 cm. That's all, thank you, and what material do you want for your blouse?

B: Oh, I'm not sure, but I want something good. Can you help me?

A: All right. This is an exceptionally good quality woolen cloth. I can assure you that the quality of the cloth is excellent.

B: That's settled.

A: Can you call in tomorrow for a fitting?

B: I'd be glad to.

>> New Words and Expressions <<

blouse	女衬衣
measurement	测量
spare no effort	尽力
beautify	美化
pattern	式样
fitting room	试衣室
a bit	一点儿
loose	宽松
waist	腰
have it styled	使定制设计
choose from	挑选
something good	好的
a bit larger	有点大
care for	喜欢（用于否定疑问句）
vary	变化
breast	胸部
decoration	装饰
main	主要的
sleeve	袖子
chest	胸围
exceptionally	特别地
woolen	羊毛的
cloth	料子
assure	保证
settled	决定了
call in	来
fitting	试样
excellent	出色的
attraction	吸引人的地方
owe much to	很大程度上由于
owe	由于
be suitable for	适合
pattern	型号，花型

Notes to the Text 1

1. Today blouses come in many colors, styles and prices. 今天女衬衫有许多颜色、款式和价格。“come in”意思是“有”。

2. This blouse is quite all right except for its color. 这件女衬衫除了颜色外都很好。

3. The tastes to colors vary from person to person. 对颜色的注重因人而异。

4. We should spare no effort to beautify our clothes. 我们不遗余力地美化我们的服装。

5. The blouse owes much to the decoration. The main attraction of the blouse is here. 这件衬衫很大程度上靠这装饰。它最吸引人的地方就在这。

6. I can assure you that the quality of the cloth is excellent. 我可以向你保证布料的质量是优异的。

7. I'll have it styled by a famous designer. 我请一位著名设计师来设计。“have something done”意思是“请某人做”，“style”意思是“按时尚设计”。

8. Can you call in tomorrow for a fitting? 你能明天来试样吗?

9. That's settled. 就这么定了。

Text 2

>> Learning How to Take Measurements 学习怎样量体

Taking measurements is a very important course in garment vocational school. It needs skill to take measurements. There is nothing like practice for acquiring skill. To learn and master this course involves practice. Taking measurements plays a decisive role in clothes. Taking measurements is one of a very important subjects we should learn and master. Whether a garment fits, to a large extent, is closely related to measurements taking. Taking measurements is a very necessary and important step in making up a garment. We should be careful when we are taking one's measurements. To learn and master the subject we should practise the skill. Skill comes only with practice. Only through constant practice, can we learn and master the skill.

Taking measurement is a very important skill,especially the garments made to order ones. Before we take measurements, we must make sure how many measurements we are going to take and what are they. We should take one's measurements according to his or her request. We must write down the measurements clearly and accurately. If you take the incorrect measurements or wrong sizes, the patterns you make are incorrect, and the garment will be not suitable for the person who wants it. So we must pay attention to it. First we must make sure what kind of garments we are going to make and what figure we are going to take measurements, and how many steps we are going to take. We should keep in mind the methods, rules, and order of taking measurements. Of course we should listen to the customer's taste and interests. We should respect their habits and needs. We should write down the sizes carefully and correctly when taking the measurements. The most important is that we should check the sizes again and again to make sure they are right and accurate.

>> New Words and Expressions <<

course	课程	related to	与……相关
involve	包括	make sure	确信
decisive	决定性的	according to	根据
master	掌握	write down	写下
practise	练习	accurately	准确地
practice	练习	incorrect	不正确
constant	不断的	figure	身材，体型
especially	尤其	keep in mind	牢记
made to order	定制的	method	方法
carefully	仔细地	order	步骤
step	步	respect	尊重
make up	制作	check	核查
fit	适合	accurate	准确的
to a large extent	很大程度	size	尺寸
closely	密切	correctly	确定地，正确地

Notes to the Text 2

1. It needs skill to take measurements. 量体需要技巧。

2. There is nothing like practice for acquiring skill. 只有通过练习才能获得技巧。

3. Taking measurements plays a decisive role on clothes. 量体对服装起着决定性作用。

4. Whether a garment fits, to a large extent, is closely related to measurements taking. decisive on a garment, especially the garments made to order ones. 一件服装是否合体，很大程度与量体密切相关，尤其是定制的服装。

5. Skill comes only with practice. Only through constant practice, can we learn and master the skill. 能力来自实践。只有通过不断地努力，我们才可以学习和掌握这项技能。

6. We should respect their habits and needs. 我们应该尊重他们的习惯和需求。

>> Substitution Drills 替换练习 <<

1. Today [blouses / fashions / sewing machines] come in many colours, styles and price.

2. I'd like to order [an overcoat / a shirt / a skirt].

3. This [blouse / dress / T-shirt] is quite all right except for its colour.

4. Many young women find the [design / dress / skirt] attractive.

5. The taste to [collars / styles / fashions] vary from person to person.

6. I like the [style / colour / pattern] but it's a bit larger for me.

Exercises to the Texts 课文练习

1. 抄写并熟记本课短语和词组
2. 翻译下列短语
   1）assure you
   2）call in
   3）the quality of the jacket
   4）too loose round the waist
   5）the taste to collars
   6）play a role
   7）make sure
   8）write down
   9）made to order
   10）keep in mind
   11）something good
   12）taste to clothes
   13）come in
   14）except for
   15）to a large extent
   16）make up
   17）that's settled
   18）have something done
   19）owe...to
   20）the main attraction of
3. 将下列短语译成英语
   1）有点大
   2）尽力
   3）定做一件大衣
   4）美化我们的生活
   5）不遗余力
   6）不断练习
   7）喜欢
   8）适合
   9）挑选
   10）有
   11）为你量体
   12）羊毛布料
   13）服装的品位
   14）因人而异
   15）美化服装
   16）使定型设计
   17）和……密切相关
   18）学好英语
   19）清楚准确地
   20）布料的质量

4. 翻译

1）量体对服装起决定性作用。

2）我想定做一件夹克衫。

3）男装师为他量了尺寸。

4）胸围太大了。

5）现在大衣有许多颜色、款式和价格。

6）适合他的未必适合我。

7）我向你保证这款大衣的质量是一流的。

8）我们应尽力学好设计。

9）量体是制作服装很重要的一步。

10）我们应按顾客的要求来给他（她）量体。

11）对服装的注重因人而异。

12）这件大衣除了颜色外都很好。

13）他手工非常灵巧。

14）量体需要技巧。

15）很多年轻女士觉得这设计很吸引人。

16）学习和掌握这门课程需要实践。

17）我请一位设计师来设计。

18）只有通过不断努力，我们才能学好服装设计。

19）这件上衣是我们设计师设计的。

20）我喜欢这件大衣，但我穿有点大。

Conversation

>> Talking About Measurements Taking 谈谈量体

Conversation 1

A: I am a vocational school student. Nowadays I'm learning how to make up a garment. Today we are having the course of measurements taking. The course has a lot of content, and we have a wide range of things to learn. I am very interested in measurements taking. But I have some difficulties in my learning. The difficulties confused me. Every time I take this course, I feel a little afraid in my heart. It is said that you have much talent for fashion design and measurements taking. With your talent and hard-working, you have made great progress in your work. Could you please tell me something about it?

B. Sure. Talent is very important but training is even better. My success is not by talent, my achievements mainly depend on continuous practice. As the old saying goes, practice makes perfect. We must be confident that we can learn it well, as long as we do not lose heart, we will at last overcome all the difficulties. Taking measurements is a very important course in garment vocational school. It needs practice. Only through continuous practice can we master it. Taking measurements plays a decisive role on clothes. When you are doing it you should listen to the teacher's instruction, according to the teacher's requirements. Remember it is necessary to be careful. In practice, the steps should be accurate, the looseness should be appropriate, and the customer's requirements should also be considered. In addition, we must establish self-confidence and believe that we can learn it well.

A: Thank you for your guidance and help. I will keep what you said in mind and keep practicing. I will share what I have learned from you with my classmates.

B: There will be a great improvement on your learning. Come back to me if you have any questions. As long as you work hard, there is hope of success. I am sure you will be a good designer.

Conversation 2

A: Last time we had a talk about measurements taking. I am deeply impressed by your profound knowledge, exquisite craftsmanship, and the method of teaching. With the help of your method, I catch up with my classmates. Now I am confident of learning it well. Whenever we are talking about the course, I can't help thinking of you. When I am in trouble, I bear your advice in mind to face up to problem with courage and get over difficulties. I told all these to my classmates and shared my learning experience with them. They are very interested in this. We are decided to adopt the new approach to learning the measurements taking.

B: I am glad to hear that. It is said you have improved your study so greatly that you could begin help the teacher explain the knowledge to the students freely in class. The accuracy of this work is so important that any negligence will cause serious consequences. Most of our mistakes are caused by carelessness. I hope you make persistent efforts, don't be proud and continue to work hard. We should always bear in mind that good results, progress and happiness are made by struggle. Now let's take the jacket as an example to learn every step of measurements taking in detail.

A: That's great. What are the specifications of a jacket? How should we take these measurements?

B: Specifications and sizes are the main basis of drawing paper patterns of a jacket. Generally speaking, we have coat length, chest width, shoulder width, sleeve length, collar size and cuff size. These measurements must be accurate. They are of great importance. Ignoring any of them will cause serious problems in drawing paper patterns and production. All the sizes must be checked repeatedly before they are used.

A: Thank you for your teaching. Our classmates will take your approach to do it. We will master the skill and will not disappoint you.

B: By the way, our company will hold a report on fashion design and clothing measurements next week. A famous designer will be invited to give a lecture. You can go and listen to it. Keep me informed of the progress of your study. I will make every effort to help you.

A: Thank you. I will.

New Words and Expressions

have the course of measurements taking	上量体课
content	内容
confuse	使迷惑
have a wide range of things to do	有很多事要做
have much talent for	某方面造诣很深
continuous practice	不断练习
play a decisive role on	起决定性的作用
looseness should be appropriate	松度适当
consider	考虑
guidance	指导
establish	建立
main basis of	主要依据
specification	规格
Ignore	忽视
master the skill	掌握技能
keep me informed of	随时通知我
be deeply impressed by	留下深刻印象
profound	渊博的
as long as	只要
exquisite craftsmanship	精湛的工艺
bear in mind	牢记
face up to	面对
be confident of	有自信
adopt	采纳
approach to	方法
accuracy	准确
negligence	疏忽
consequence	后果
be made by struggle	靠奋斗得来的
accurate	精确的
be checked repeatedly	反复核对
disappoint sb.	使某人失望
progress of your study	你学习进展情况

Reading Practice ❶

Introducing My Workroom 介绍我的工作间

I have a workroom in my apartment. It is about twelve square meters. There is a cutting table and a working table in my workroom. I have two sewing machines by the wall, one is made in China and the other is made in America. The sewing machines work well. I clean and oil the machines every day. I think sewing machine has become an essential part of my daily life. Usually I am busy with sewing on Saturdays and Sundays. I design and sew many coats, dresses, trousers, jackets and shirts for my family with the sewing machines. When I see my family wear the new garments, I get a sense of achievement. I stick to practicing designing, stitching and sewing, and I will get great achievement sooner or later. I am sure that I will become a good designer and tailor in the future.

I have all the tools for sewing in my workroom, including two pairs of scissors, a tape measure, three rulers and some tailor's chalks. On the working table there is a large pressing board. There is a steam iron and a piece of pressing cloth on the board. I press and iron my garments with the steam iron.

I draw design pictures at the table and I am interested in painting because it allows me to express my feelings about beauty.

New Words and Expressions

square meters	平方米	iron	熨烫
oil	给……加油	press	压烫
be busy doing sth.	忙于做某事	pressing cloth	烫布
press board	烫板	scissors	剪刀
steam iron	蒸汽熨斗		

Reading Practice ❷

My Workroom 我的工作间

Here is my workroom. It is not very large. It is about ten square meters. I have a cutting table, two sewing machines, a steam iron, some useful tools and other equipment. In addition, I have the tools of my trade. The sewing machines, tools and other equipment work well. I clean and oil the sewing machines every day. On the desk there are some books about clothes and design. I have many design magazines on the desk. These books and magazines have a good influence on me. When I meet with some problems I turn to my teacher for help, and sometimes I read the books and magazines to look up answers. I am keen on reading because reading can bring me knowledge and happiness.

Every Saturday and Sunday I design my clothes in my workroom. I draw pictures, make patterns of coats, trousers, dresses, jackets and shirts for myself and my family. I cut out these clothes according to the patterns and baste them for a fitting. If some places and positions don't fit, I will make some alterations to them. After alterations, I try them on for the final fitting. When I am sure that everything is all right. I stitch, sew, and press the clothes. The garment I design is a perfectly fit. I choose best material and my workmanship is perfect. I am pleased with my work.

New Words and Expressions

in addition	除此之外	fitting	试样
trade	行业	mark	标出
magazine	杂志	make some alterations	做些改动
have a good influence on	起……有好作用	try...on	试穿
turn to sb. for help	向某人求助	the final fitting	最后试样
look up	找	perfectly	完美地
be keen on	喜爱	workmanship	工艺
bring	带给	perfect	完美的
baste	粗缝	be pleased with	满意

Lesson Seven

Text ❶

>> The Garment Packing 服装包装

A: Good morning, Mr. Smith. Glad to see you.

B: Good morning, Mr. Green. Glad to see you, too.

A: Sit down, please. Have a cup of coffee.

B: Thank you.

A: Can I help you?

B: I'm interested in your products. I looked at the goods on display. The style is very novel and the quality is perfect, but I am not sure about the packing of your products. Could you please tell something about the packing? You know we pay much attention to packing.

A: Thank you for your interest in our products. Please be assured that our packing is perfect. We believe that packing can create a high perceived value for goods and products. We pack the shirts in boxes of half a dozen each, 16 boxes to the carton.

B: Can you tell me the dimensions of the carton?

A: Yes, they are 82cm long, 77 cm wide and 51 cm high.

B: What's the volume?

A: It's about 0.32 cubic meter.

B: And the weight?

A: The gross weight is 48.5 kg, the net 47 kg, you see, the tare weighs is 1.5 kg only.

B: By the way, what strapping do you use for the cartons?

A: Plastic strapping.

B: That's good. Plastic strapping is light, strong and handy.

A: You said it. All our efforts are for the customers.

B: Thank you.

A: Don't mention it.

B: Good-bye, Mr. Green.

A: Good-bye, Mr. Smith. I'm glad to have seen you..

New Words and Expressions

believe	相信	volume	体积
on display	陈列	cubic	立方
perfect	完美的	gross	毛重
packing	包装	net	净
novel	新颖	tare	皮重
goods	货物	strapping	打包带
please be assured	请放心	plastic	塑料
create	产生，创造	handy	轻便
perceived value	视觉价值	you said it	同意，赞成你的意见
dozen	（一）打（12个）	effort	努力
carton	纸箱	weight	重量
dimension	尺寸		

Notes to the Text 1

1. Packing can create a high perceived value for goods and products. 包装能对货物和产品产生很抢眼的视觉效果。

2. All our efforts are for the customers. 我们所做的努力都是为了顾客。

3. Glad to have seen you. 见到你很高兴（分手时说）。

Text 2

Packing 包装

Packing is a very important part in clothing industry. We pay much attention to the quality of the products. We must also pay much attention to the package and packing. Proper packing is almost as important as the quality of the product itself. Many people pay attention to the quality of the products. But they sometimes ignore the packing and fail to pack the goods properly. They lost to other products in the market. Good packaging can catch the customers' eyes. The customers often see the package and packing as its contents as a whole. Good package gives people with beautiful enjoyment. Good package gives people shine at the sight of the products. So beautiful package and packing add value to the products. Many people, especially young ladies focus their attention on the package and packing of products. So many companies and firms pay much attention to their package and packing. Their beautiful packages and packing attract many customers.

Good packaging makes the product pleasing to the eye. Proper packaging makes products more competitive. When we sell our garments, we should pay special attention to the packaging and packing. Goods that are attractively packaged sell more quickly. If you want to make the products sell well, you must make your products more attractive than others'.

New Words and Expressions

package	包装物
proper	适当的，恰当的
ignore	忽视
fail to	未能
properly	适当地
catch the customers' eyes	吸引顾客的眼球
as...as	像……一样
content	内容
shine	发光
at the sight of	一看到
add	增加
focus on	把注意力放在
make...please to the eye	使……令人赏心悦目
competitive	竞争力
firm	商号

Notes to the Text 2

1. Proper packing is almost as important as the quality of the product itself. 合适的包装和产品的质量一样重要。

2. Good package gives people shine at the sight of the products. 好的包装让人们看到产品就眼前一亮。

3. Goods that are attractively packaged sell more quickly. 包装吸引人的货物销售得更快。

4. Good packaging makes the product pleasing to the eye. 良好的包装使产品赏心悦目。

5. Proper packaging makes products more competitive. 恰当的包装使产品更具有竞争力。

Substitution Drills 替换练习

1. Please tell me something about the packing / measurement / coat / trousers.

2. We pack the shirts / pants / shorts in boxes of half a dozen each.

3. Can you tell me the dimensions of the carton / box / desk / case.

4. Glad / Nice to have seen you.

5. All our efforts are for the [customers / students / staff members].

6. What's the [volume / length / width]?

Exercises to the Texts 课文练习

1. 抄写并熟记本课短语和词组
2. 翻译下列短语
   1）be interested in
   2）cubic meter
   3）half a dozen
   4）the dimension of the box
   5）perceived value
   6）please be assured
   7）all our efforts
   8）at the sight of
   9）focus on
   10）as...as
   11）gross weight
   12）net weight
   13）plastic stripping
   14）the volume of the box
   15）pay special attention to
   16）attract customers
   17）add value to
   18）fail to do sth.
   19）the volume of a box
   20）the dimensions of a box
3. 将下列短语译成英语
   1）我们所有的努力
   2）轻便结实
   3）毛重
   4）桌子的尺寸
   5）宽度
   6）纸箱的尺寸
   7）产生抢眼的视觉效果
   8）陈列
   9）完美的
   10）一见到
   11）吸引顾客的眼球
   12）一些有关熨烫的事项
   13）创造价值
   14）注意
   15）增加
   16）更吸引人
   17）未能
   18）轻便的
   19）尤其
   20）畅销

4. 翻译

1）我们所做的努力都是为了服装的质量。
2）衣领能让服装产生很抢眼的视觉效果。
3）我对你们的衬衫很感兴趣。
4）你能告诉我这盒子的尺寸吗?
5）塑料打包带结实轻便。
6）包装对货物和产品很重要。
7）我们的短裤每半打装一纸盒。
8）纸箱的体积是0.3立方米。
9）我们出售服装时要特别注意包装。
10）合适的包装和服装的质量一样重要。
11）我们所有的努力都是为了学生。
12）让我们谈谈熨烫的事项。
13）你考虑得很周到。
14）长度是多少?
15）他们忽视了包装未能恰当地包装货物。
16）我们的衬衫每半打装一盒子。
17）我对你们的夹克感兴趣。
18）请放心我们的质量是一流的。
19）包装对我们很重要。
20）良好的包装能吸引顾客的眼球。

Conversation

>> A Dialogue About Packing 谈论包装

A: I am a businessman. I'm interested in importing garments from you.

B: Thank you, Mr. A. I'm a sales manager of the company. Sit down, please.

A: I want to know something about your packing. Let's start discussing the packing, shall we?

B: OK, you know we have good ways of packing blouses. You see we use polythene wrapper for each article.

A: Very good. How beautiful it is! The goods must not only be good value but also attractive. What about outer packing?

B: We'll pack them in cartons.

A: I'm afraid cartons are not strong enough for such a heavy load.

B: The cartons are light and easy to handle. The cartons with plastic sheets are waterproof.

A: Can you tell me the dimensions of the cartons?

B: Yes, they are 70 cm long, 60 cm wide and 40 cm high. The volume is 0.168 cubic meter. The gross weight is about 44 kg. The net weight is 42 kg. You see, the tare weight is only 2 kg.

A: What strapping do you use for the cartons?

B: Plastic strapping. Plastic strapping is light, strong and handy. All our efforts are for the customers. Goods for export are specially packed. Our packing is usually suitable for long distance transportation. Some of the goods are packed in seaworthy wooden cases. Do you have any suggestions to make? We will take your suggestion into consideration.

A: It is very considerate of you.

B: We assure you of our cooperation and listen to your further comments.

Words and Expressions

import from	从……进口	gross weight	毛重
sales manager	销售经理	tare weight	皮重
polyether wrapper	聚乙烯包装袋	export	出口
article	商品	be specially packed	特别包装
good value	货好	long distance transportation	长途运输
look attractive	看上去吸引人	seaworthy wooden case	适用航海的木箱
outer packing	外包装	it is very considerate	某人考虑得很周到
easy to handle	容易搬运	assure sb. of sth.	使某人相信某事
plastic sheets	塑料纸	cooperation	合作
waterproof	防水	further comments	进一步的意见
make some suggestions	提建议	consideration	考虑
cubic meter	立方米		

Reading Practice ❶

Clothes Packing 服装包装

Packaging is a very important working procedure in a garment factory. Packaging is an important part of garment production. Many people have neglected this before. Nowadays more and more people realize the importance of packaging. They are making every effort to improve the packing. They are making products more beautiful and attractive. They improve the design, the color and the materials.

They use the most advanced packaging technology. We used to pack garments in wooden cases. Now we pack them in cartons. Cartons are light and handy. There are polybags in them. We use strong cartons and double straps.

We print marks on both sides of a carton. We print goods number, goods name, quantity, weight, contract number and so on. Sometimes we print "This side up" on the cartons.

This kind of carton ensure against damage. We should pack the garment carefully. Don't crumple our garments. If we pack our garments well. The garments will look good when we sell them. Whether the appearance of clothing is smooth or not will affect people's overall view of clothing. Now we have done business with many countries and regions all over the world. We must pay much attention to the safety of outer packaging.

New Words and Expressions

procedure	工序	polybag	胶袋
print	印	ensure	保证
mark	唛头，记号	against damage	抗损
neglect	忽视	damage	损坏
advanced	先进的	crumple	压皱
contract	合同	realize	意识到
appearance	外观	the importance of	重要性
smooth	平整的	make every effort	尽力
overall view	总体看法	this side up	此端向上
outer	外面的，外部的	affect	影响

Reading Practice 2

Our Clothes Packing 我们服装包装

Packing is a very important work process in making up clothes. Package is especially important for clothes in the market. Many companies packed the clothes in eye-catching packing materials. They want their goods to be more popular in the market. They want their products to sell more and more quickly. They want their products dominate the market.

Nowadays more and more customers especially young people have realized the importance of package and packing. The managers and designers in our country pay much attention to the package and packing. They make every effort to improve packaging materials, color and design. They achieved their goals. Now the products in our market are colorful. The products attract people. Our products are more attractive to customers. Our products enjoy popularity in the market both at home and abroad. People believe great achievement can be made only through hard work. Our woolen, cotton and silk clothes sell best in the market. We have a good market in many countries and regions. There is a good market for our clothes.

We usually use plastic bags for each article and ten dozen or twenty dozen to one carton. The cartons are strong enough for the load, but they are light and easy to handle. Sometimes we use wooden cases or customer specified package. We use plastic strapping for cartons. Plastic strapping is light, strong and handy. All our efforts are for the customers.

New Words and Expressions

process	过程，工序	carton	纸箱
eye-catching	吸引眼光	specified	特别的
dominate	支配	achieve	取得
woolen	羊毛的	enjoy popularity	享有名望
regions	地区	achievement	成就
plastic bag	塑料袋	goal	目标
article	商品	at home and abroad	国内外

Lesson Eight

Text 1

>> The Garment Fair
服装交易会

Here we are in the Garment Hall of the Export Commodities Fair. The hall is decorated with colorful flags and balloons. A great variety of new and traditional garments are on display. Recent years, we have established business relations with more and more companies and firms of all parts of the world. Every year a large number of merchants and businessmen abroad come to visit the fair. The products displayed attract thousands of visitors, guests, merchants and businessmen from home and abroad to the place. Some of them have placed their orders with our companies and factories for clothes. Many of them already have intention to place large orders with us for clothes.

A guide is showing a number of businessmen and guests round the hall. Some businessmen and guests are from foreign countries, and some are from all parts of our country. They are looking at the exhibits and pictures with great interest. What interests the foreign visitors is how the Chinese clothes attract customers' attention. By the window, many overseas Chinese are taking photographs. In front of the fashion show, a great many foreign merchants are chatting about the color and styles of the dresses and Chinese gowns. And many foreign businessmen and merchants are eager to get firm offers from our representatives. In the discussion rooms, a lot of businessmen are having talks with our representatives. Some are making inquires, some are talking about the prices, some are placing orders, and some are drawing up contracts and agreements. They are doing business on the spot. Some order the goods on the spot.

The exhibition presents a picture of prosperity in our export clothing, and it reflects the new look in the field of the garment industry of our country. We can get a lot of information from the exhibition and gather suppliers.

New Words and Expressions

hall	大厅	balloon	气球
be decorated with	用……布置	exhibit	展品
commodity	商品	overseas Chinese	海外华人
fair	交易会	take photograph	拍照
variety	品种繁多	chat	聊天
display	陈列	Chinese gown	旗袍
establish	建立	representative	代表
business relations	贸易关系	make inquire	咨询，询价
have intention to do	想做某事	place an order	订货
dealer	客户	agreement	协定
merchant abroad	海外商人	on the spot	当场
guide	导向者	present	展现
with great interest	怀着很大兴趣	prosperity	繁荣
firm offer	实盘报价	reflect	反映
attract	吸引	gather supplier	组织货源

Notes to the Text 1

1. Recent years, we have established business relations with dealers of all parts of the world. 最近几年我们已和世界各地的商人建立了贸易关系。

2. have intention to do sth. 有意向做某事。例如：They have intention to buy your products. 他们有意向购买你们的产品。

3. They are looking at the exhibits and pictures with great interest. 他们正怀着极大地兴趣观看展品和图片。

4. place an order with sb. for sth. 向某人订购某物。例如：He placed an order with the businessman for ten coats. 他向那位商人订购了十件上衣。

5. The exhibition presents a picture of prosperity in our expert garments, and it reflects the new look in the field of the garment industry of our country. 展会呈现出我国出口服装的繁荣景象，并反映了我国服装工业领域的新面貌。

6. What interests the foreign visitors is how the Chinese clothes attract customers' attention. 外国参观者们感兴趣的是中国服装是如何吸引消费者的注意的。

Text 2

>> Attending a Garment Fair 参加服装交易会

Here we are in an exhibition hall. It is a garment fair. We are invited to attend the garment fair. The city government holds a fair once a year. Now let's go into the Garment Hall of the Export Commodities Fair. You can see a great variety of new and traditional garments are on display. They are beautiful and nice. Many of them have Chinese elements, such as panda, dragon, phoenix, Chinese characters and Great Wall. Visitors and businessmen are very interested in these logos. The clothes displayed attract thousands of visitors and businessmen from home and abroad to the place. Recent years, we have built up and established business relations with dealers of all parts of the world. We have been dealing with many businessmen and merchants from all over the world. They have special interest in our garments. They are going to place orders with us for garments.

Here is our garment hall. A lot of visitors and foreign businessmen are looking at our products and pictures with great interest. At the table, a few foreign merchants are talking about the color and styles of the clothes and Chinese gowns. Some Chinese designers and staff members are having business talks with foreign businessmen in English. The designers of all companies are trying their best to cater for all their tastes and needs. Some merchants are making inquires, and some are placing orders. Some businessmen and our staff members are drawing up contracts. Many businessmen are signing the contract on the spot.

New Words and Expressions

attend	参加	build up	建立
commodity	商品	relation	关系
element	元素	dealer	商人
panda	熊猫	deal with	生意往来
such as	例如	cater for	满足
dragon	龙	have special interest in	对……特别感兴趣
Chinese character	中国书法	staff member	工作人员
phoenix	凤凰	have business talks with	和……贸易会谈
logo	标识，标志	merchant	商人
establish	建立	sign contract	签合同

Notes to the Text 2

1. The designers of all companies are trying their best to cater for all their tastes and needs. 公司所有的设计人员尽力满足他们的品位和需求。

2. Many of them have Chinese elements, such as panda, dragon, phoenix, and Great Wall. 它们中许多都含有中国元素，例如熊猫、龙、凤凰和长城。

3. We have been dealing with many businessmen and merchants from all over the world. 我们一直在和世界各地的商人进行贸易往来。

4. Some businessmen and our staff members are drawing up contracts. 一些商人正在和我们的工作人员草拟合同。

5. Many businessmen are signing the contract on the spot. 许多商人当场签了合同。

Substitution Drills 替换练习

1. Here we are in the Garment Hall of the fair / exhibition / company .

2. A great variety of new and traditional [dresses / fashions / skirts / suits] are on display.

3. A lot of [businessmen / guests / friends] are having talks with our representatives.

4. They attract thousands of [visitors / businessmen / foreigners] to the place.

5. They are looking at the [clothes / exhibits / fashions / dresses] with great interest.

6. A great variety of [new and traditional garments / sewing machines / steam irons] are on display.

Exercises to the Texts 课文练习

1. 抄写熟记本课单词和词组
2. 翻译下列短语
   1）export commodities fair
   2）on display
   3）establish business relations with
   4）merchants abroad
   5）a great variety
   6）traditional garments
   7）a lot of businessmen
   8）firm offer
   9）deal with
   10）build up
   11）present a picture of prosperity
   12）take photographs

13）the new look
14）with great interest
15）place order
16）on the spot
17）try one's best
18）cater for
19）attract one's attention
20）have intention to do

3. 将下列短语译成英语

1）海外华人
2）当场
3）吸引来访者
4）订货
5）向某人订购
6）贸易关系
7）怀着很大兴趣
8）组织货源
9）咨询
10）草拟合同
11）服装工业领域
12）旗袍
13）我们的代表
14）在……领域
15）签合同
16）展示
17）服装交易会
18）进入
19）国内外
20）全世界

4. 翻译

1）各种各样新颖及传统服装正在交易会服装厅展出。
2）各种新式衬衫吸引着成千的参观者。
3）玛丽怀着很大兴趣在看那本设计书。
4）几位外商正在观看展品。
5）他们当场订了货。
6）那个商人正在询价。
7）展览会显示出我国服装的繁荣景象。
8）我们的代表正在带领大批外商参观工厂。
9）我们尽力迎合顾客的品位和需求。
10）我们应邀去参观服装交易会。
11）我们有意向购买他们的夹克。
12）我们学校呈现出新面貌。
13）他们很想得到实盘报价。
14）陈列的服装吸引了很多商人。
15）外商对我们的中国元素感兴趣。
16）他们和许多国家建立了贸易关系。
17）我们怀着很大的兴趣观看展品。
18）他们正在草拟合同。
19）我们向那家公司订购了很多大衣。
20）我们一直和他们有生意往来。

Conversation

>> Talking About the Garment Fair
谈论本次服装交易会

A: It is said that you attended the garment fair held in our city. Could you please tell me something about it?

B: Sure. Last week I received an invitation from the Fashion Association. They invited me to attend the garment fair held in the Exhibition Hall. It is the biggest exhibition in our country. There are a variety of modern and traditional clothes displayed there. Many of them are designed by our top designers. Some of them contain a lot of Chinese Dream elements. For designers，China and its rich culture are the inspiration. They come up with creative ideas and clothes to improve people's life. The fair attracted a large crowd. The fair was crowded with designers, businessmen, merchants clothing staff members and many visitors from home and abroad. Many companies, factories and enterprises participated in the garment fair in order to gain greater share in the international and domestic markets. In addition to clothing products, many enterprises and scientific research institutions have also brought a lot of new materials and mechanical equipment. In the past five days the exhibition has received a positive response from visitors. The garment fair is worth seeing and visiting. You should go to see it, if you miss it, you will regret it.

A: I'm excited about what you said. I wish I could visit it right now. I am interested in the processing technology and methods of clothing. Do you have any new information in this field? I have been looking for the information about them. In particular, I would like to see their innovative ideas and innovative products. If I want to get some information on high technology, what companies and institutions do you suggest I should contact?

B: Every day they hold lectures on fashion trends, domestic and international marketing situation. The exhibition also invited many professionals to discuss relevant professional issues. The companies, factories and scientific research institutions each have their own showing room. You can go directly to their exhibition hall or ask the exhibition staff for help. I'm sure they will provide you with the detailed information. This fair is different from previous ones. By visiting this fair, you will surely get creative inspiration. It is very hard for those who haven't been to the fair to describe its technological content.

A: Thank you for your help. It is very kind of you to give me such a detailed introduction.

New Words and Expressions

Fashion Association	服装协会	innovative idea	创新理念
top designer	顶尖设计师	high technology	高科技
inspiration	灵感	contact	联系
come up with	提出	fashion trends	流行趋势
creative idea	创新理念	marketing situation	营销状况
merchant	商人	professionals	专业人员
enterprise	企业	professional	专业的
participate in	参加	relevant	相关的
gain greater share	获得更大的份额	issue	问题
scientific research institution	科研单位	provide sb. with sth.	向某人提供某物
mechanical equipment	机械设备	detailed information	详细信息
positive response	积极响应	previous	以前的
regret	后悔	creative inspiration	创造灵感
processing technology	加工工艺	describe	描述
in particular	尤其特别	technological content	技术含量

Reading Practice 1

Visiting a Garment Fair
参观服装交易会

Last week, I went to the garment fair. At the fair a great many of fashionable garments, pictures and photos were on display. They attracted hundreds and thousands of visitors to the place. I looked at the exhibits and pictures with great interest. There was a fashion show at the fair, too. The overcoats and dresses on show were very beautiful and well-pressed. Many people enjoyed the show. There were a lot of vocational school students. Many foreign businessmen from all parts of the world visited the fair. There were many staff members received them. They talked with the designers and representatives of the fair. The businessmen and merchants asked many questions and made some suggestions. The designers and representatives accepted the suggestions and answered all their questions with complete assurance. What a wonderful fair it was. I enjoyed every minute of it. I'm proud of our products. I'm proud of our country. I got a great deal of enjoyment from seeing the exhibits and pictures.

There were many discussion rooms near the fair. In the rooms, many representatives from garment factories and garment companies did business with the businessmen all over the world. Some were talking about the patterns, some were signing contracts, others were bargaining over the prices. How busy they were!

New Words and Expressions

fashionable	时髦的
well-pressed	挺括的
enjoy every minute	从头到尾都欣赏
be proud of	为……而骄傲
with complete assurance	满怀信心地
discussion room	洽谈室
patterns	式样，样板，纸样
sign contract	签合同
bargain over the price	讨价还价
a great deal	许多
enjoyment	愉快，乐趣

Reading Practice 2

A Garment Fair
一场服装交易会

A new garment fair has been opened up near the center of the city. It is one of the most important garment fairs in our city. The clothing fair is held once a year. It gives a new look of our clothing to the world. It shows the world our new look of foreign trade. We try to keep up with advanced countries in clothing. The garment fair attracts great attention of the people. The garment fair attracts many people to visit. Many of them are from all parts of the world. We have many new products and traditional garments are on display at the fair. To cater to the needs of the public and customers, many clothes with Chines elements are designed. These products consist of Chinese traditional culture. Based on some Chinese elements, the clothes displayed here win great popularity and visitors and businessmen.The traditional and new clothes on display attract many foreign visitors and hundreds of visitors to the place. Our garments enjoy great popularity and a very high repute at the fair. Many foreign businessmen talk with our designers and representatives of the fair.

We have many discussion rooms near the fair. In the discussion rooms, our designers and staff are having a heated and lively discussion about the new clothes with a foreign businessman. They discuss all the technical details and a series of quality requirements. They draw up the contract terms in detail. They want to reach an agreement on the contract. The result of business talk is satisfactory. They will have a further discussion and sign the contract tomorrow.

New Words and Expressions

hold	举办
advanced country	发达
keep up with	赶上
our new look of	我们……的新面貌
foreign trade	外贸
consist of	包含
based on	基于
attract great attention of	引起高度注意
open up	开幕
merchant	商人
enjoy great popularity	备受欢迎
high repute	很高的声誉
technical details	技术细节
heated	热烈的
lively	活跃的
a series of	一系列
quality requirement	质量要求
draw up	草拟
term	条款
in detail	详细地
reach an agreement	达成协议
result	结果
further discussion	进一步讨论
satisfactory	满意
sign	签署

Lesson Nine

Text ❶

A Dialogue in the Cutting Workshop
裁剪车间的对话

A: Are you Master Wang in the cutting workshop?

B: Yes, you must be the new workers. I'm very glad to see you. Welcome to our factory.

A: Yes, you are right. We just graduate from the vocational school. We new workers are not familiar with this kind of work. We are very worried. We are eager to become skilled workers soon.

B: Don't worry. I shall help you with your work. I give the assurance that I will readily offer you more help. It is my duty to help you. Remember, once you enter the workshop, you should follow the masters' and coach's instructions. You should pay close attention to safety. You must keep in mind that while working, the slightest carelessness will cause an accident. Safety is the first. Look at the sewing machine. It was invented by our engineers. It is a highly efficient sewing machine.

A: It's very kind of you. What are you going to do now?

B: Today I'm going to cut out a new jacket. It's the latest kind this year. The design will be of interest to customers.

A: Oh, how are you going to cut out a jacket? Is it very easy?

B: No, it is not easy. It is difficult to cut out a jacket well. First, we should take measurements. Now I'll tell you the way to take measurements. There are many parts we should take the measurements.

A: How many parts shall we measure?

B: There are many parts, such as length, chest, shoulder, sleeve length, armhole, cuff, collar and the bottom of a jacket.

A: Then what shall we do after that?

B: Then you can draw the paper patterns according to the measurements, cut out the paper patterns, check all the sizes are right and put the patterns on the materials.

A: Now can we cut out the jacket?

B: Before cutting out you must check all the measurements and pieces. Make sure that all the measurements and pieces are right. When cutting out garments you should take measures to insure your safety and others' safety. Keep in mind that a moment of carelessness will bring about a serious accident. Then you can cut out the jacket. Do you catch it?

A: Yes, we shall keep it in our minds. Thank you.

New Words and Expressions

master	师傅	armhole	袖窿
graduate	毕业	readily	乐意地
familiar	熟悉的	remember	记住
follow	听从	offer	提供
coach	实习老师	cuff	袖口，袖头
invent	发明	highly efficient	效率高的
be eager to	迫切地	bottom	底边
give the assurance	保证	draw the patterns	制图，制板
instruction	指导	according to	根据
it's one's duty to do	做……是某人的责任	check	检查
latest	最新的	pieces	衣片

be of interest to	感兴趣	make sure	确信
take measurements	量体	insure	确保
measure	测量	take measures to do sth.	采取措施做某事
length	长	safety	安全
shoulder	肩	Do you catch it?	听明白了吗?
assurance	保证	slightest carelessness	轻微的疏忽
carelessness	粗心	keep in mind	牢记
accident	事故	moment	瞬间
bring about	造成，引起，导致	serious	严重的

Notes to the Text 1

1. I give the assurance that I will readily offer you more help. It is my duty to help you. 我保证我会很乐意地帮助你们。帮助你们是我的责任。

2. Once you enter the workshop, you should follow the masters' and coach's instructions. 一旦你们进入车间要听从师傅们和实习老师的话。

3. Make sure that all the measurements and pieces are right. 确保所有的尺寸和部件都正确。

4. When cutting out garments you should take measures to insure your safety and others' safety. 在裁剪衣服的时候要采取措施保证自己和别人的安全。

5. Keep in mind that a moment of carelessness will bring about a serious accident. 记住哪怕是一瞬间的粗心也会造成严重的事故。

6. We are eager to become skilled workers soon. 我们想很快成为技术工人。

Text 2

Cutting Out a Shirt 裁剪衬衫

Peter is a new worker. He just graduated from a vocational school. He is keen on clothes. He has a strong interest in clothes. He is eager to become a skilled worker. He wants to know something about cutting out garments. Now I tell the new worker something about cutting out a shirt. I put emphasis again and again on the importance of observing safety rules. First once he enters the workshop he must follow the master worker's and coach's instruction. He must pay attention to the safety. In a word, we can't neglect the safety problem when we are working in the workshop.

Before we cut out a shirt, we should take measurements. We have many parts to take measurements. Let's take this shirt as an example, such as length, chest, shoulder, sleeve length, armhole, cuff, collar and so on. After taking measurements, we draw patterns and make paper patterns according to the measurements. After we make paper patterns we must check every size of the patterns and draw grain lines and position marks on each pattern, then cut out the patterns. We lay out the patterns on the materials, then we check the numbers of the pieces and sizes. Make sure that all the measurements and pieces are right. The right side of every piece is correct. The grain line of every piece is parallel with the fabric warp. And the position marks on the piece are all right. Remember every piece of patterns must be put in right position. Any carelessness will bring about bad result. When we are sure that everything is all right, we can cut out the shirt.

>> New Words and Expressions <<

tell somebody something about 告诉某人某事
put emphasis on 强调
again and again 再三
the importance of ……的重要性
in a word 从而言之
take...as an example 以……为例
neglect 忽视
lay out 排料
grain line 直丝缕
make sure 确保
right side 正面
be parallel with 与……平行
fabric 布料，织物
wrap 经线
position mark 定位标识

Notes to the Text 2

1. be keen on sth. 对……很有兴趣，喜爱，喜欢。be keen on doing sth. 喜欢做某事。

2. have a strong interest in sth. 对某事有强烈的兴趣。

3. be eager to become... 渴望成为……

4. put emphasis again and again on 再三强调，反复强调。

Substitution Drills 替换练习

1. I'm going to wear my [blue suit / jacket / skirt and blouse / new dress] today.

2. It's difficult to [cut out / design / make up] a jacket well.

3. I'll tell you the way to [take measures / press / cut out].

4. Make sure that [all the tools / all the materials / measurements] are right.

5. We should follow the [teachers' / masters' / coaches'] instructions.

6. How many parts shall we [measure / divide into / cut out]?

Exercises to the Texts 课文练习

1. 抄写熟记本课单词和词组
2. 翻译下列短语

1）cut out a pair of trousers
2）according to
3）keep in mind
4）sleeve length
5）grain line
6）position mark
7）lay out
8）armhole
9）fabric
10）be parallel with
11）our duty
12）do you catch it
13）take measures to do sth
14）the bottom of a coat
15）make sure
16）master worker
17）be familiar with
18）sleeve length
19）cuff
20）the bottom

3. 将下列短语译成英语

1）裁剪车间
2）量体方法
3）袖窿
4）衣长
5）毕业于
6）帮助某人某事
7）保证
8）乐意帮助你
9）做某事是我的责任
10）进入车间
11）画在衣料上
12）确保尺寸正确
13）裁剪一件上衣
14）熨烫方法
15）牢记
16）密切注意
17）根据
18）核查尺寸
19）正面
20）经线

4. 翻译

1）我们应量哪几个部位?
2）我们应采取措施保证安全。
3）他打算做件大衣，这是最新的款式。
4）量体时，先量胸围。
5）先在衣料上画样，然后裁剪。
6）裁剪前，先检查一遍。
7）我们应听实习老师的话。
8）在车间里要特别注意安全。
9）一旦进入车间要听师傅的话。
10）外商对我们的大衣感兴趣。
11）我们和他们建立了贸易关系。
12）现在我们进入了缝纫车间。
13）确保所有的尺寸准确。
14）他们当场签订了合同。

15）交易会呈现了我国出口服装的繁荣景象。

16）她在询价。

17）外商怀着很大兴趣参观了展会。

18）为了迎合顾客的需要设计师设计了这款夹克。

19）总而言之，我们不能忽视安全问题。

20）帮助你是我的责任。

Conversation

>> A Dialogue in a Garment Factory 服装厂的对话

Wan Ming was from Nanjing. He was a tailor. He was at school before he became a tailor. He was interested in clothing. Last week he visited a garment factory. Lao Li, the head of the garment factory met him at the gate of the factory.

L: How do you do? Xiao Wang. Welcome to our factory.

W: How do you do? I am glad to see you.

L: Let me show you around our factory. Here is the design room. It is one of the most important places in our factory. Here is our technical center. All the fashionable and latest designs come from here. All our creative ideas come from here. The designers and workers with the creative efforts and professional skills to design and make samples for our factory and company.

W: What's the function of it?

Li: Its function is to design and make sample garments that will look good, fit well and be practical to produce in bulk. These are our designers. Their job is to select fabrics and trimmings to create commercial design suitable for the market. Those are pattern cutters. The pattern cutter's job is to translate the designer's ideas with accuracy, and so the pattern cutters are very important partners in the production of the samples.

W: What are the workers over there?

Li: They are the sample machinists of the room. Their job is to assemble garments as efficiently as possible. They have experience in making up of samples. They make every effort to do their job well. We need everybody's wisdom on the project in order to design fashionable garments. Now let's have a rest. Next we are going to visit the other places of the factory.

W: It's very kind of you to show me around the factory.

New Words and Expressions

head of the factory	厂长	translate	表达
function	作用	designer's idea	设计师的思想
sample garment	样衣	with accuracy	精确地
fit well	非常适合	look good	看起来漂亮
be practical to produce	适合生产的	partner	伙伴
in bulk	大批量	machinist	缝纫工
select fabric	选料	assemble	制作
trimmings	辅料	have experience in	有……经验
commercial design	有商业价值的设计	efficiently	有效地
pattern cutter	样板工	project	工程
professional	专业的	effort	努力
project	项目，工作	wisdom	智慧
in order to	为了	creative effort	创造性的努力

Reading Practice ❶

Visiting a Cutting Workshop
参观裁剪车间

Let me show you around our cutting workshop. We cut out shirts, skirts, jackets coats and overcoats here. This is the first step in the production of clothing. Safety production is particularly important in this workshop. We cannot be too careful when we are cutting out the clothes. No one can ignore it. Many accidents are caused by neglecting safety in production.

Now here we are in the cutting workshop. This is our cutting workshop. It is one of the busiest workshops in our factory. There are five cutting tables in the workshop. You can see many workers working there. Some of them are taking care of the machines and equipment, some of them are blocking the fabrics. It is very important to block the fabrics before cutting out garments. We ensure the correct hang to a garment after it is stitched and sewed. Some of them are drawing patterns on the materials, some of them are checking the pattern numbers, grain lines and some of them are spreading the cloth and materials onto the cutting table, the others are cutting out the front pieces, back pieces, sleeves, facings, collars and so on. How busy they are! Everyone is going all out to fulfill the task. We can see a picture of prosperity in our garment factory.

New Words and Expressions

cut out	裁剪
the first step	第一步
ignore	忽视
front piece	前片
back piece	后片
particularly	尤其
block	验布
neglect	忽视
ensure	确保
spread	铺料
facing	过面
go all out	全力以赴
fulfill	完成
in the production of	生产……
safety production	安全生产
cannot...too...	越……越好
be caused by	由……引起
facing	门襟
correct	正确的
hang	垂势
prosperity	繁荣

Reading Practice 2

A Cutting Workshop 裁剪车间

This is our cutting workshop. We cut out all kinds clothes and materials here. It is one of the largest workshops in our factory. Let me make a brief introduction of our cutting workshop and show you around the workshop. This is one of the busiest workshops in our factory. There are ten cutting tables and other working tables in it. Every day we cut out hundreds and thousands of all kinds of clothes. You can see many workers and technical personnel working there. Some of them are preparing and blocking the fabrics. Some of them are laying out the materials onto the working tables and some of them are putting on the patterns and checking the marks and grain lines of the materials carefully. We must carefully examine each of these processes before we cut out clothes. Some of them are running the machines. They are all busy doing their jobs. They work hard and they are careful enough to check up every detail. Everyone tries his best to do the work well and fulfill the task. The skilled workers are qualified for the position. From the production of the cutting workshop and the spirit of our workers, you can see a picture of prosperity of our factory. Our factory takes on a new look everywhere. This is one of the reasons why we have achieved so good results in production recent years.

>> New Words and Expressions <<

brief introduction	简单介绍	check up	核查
technical personnel	技术人员	fulfill the task	完成任务
lay out	铺料	be qualified for	合格
run the machine	运转机器	position	位置
examine	检查	spirit	精神
process	工序	reason	原因
achieve	取得	take on a new look	呈现新面貌

Lesson Ten

Text 1

>> Placing an Order for a New Jacket
定做新套装

Last week I went shopping with my friends, I went to a famous garment store to buy a jacket for myself. But the color and style I liked was not available. The shop assistant told me that they had been sold out. The jacket was currently out of stock. The next stock would be in a month and he would inform me once it was available. One of my friends told me that the style seemed to be out of date. It is not the fashion of the year. He suggested that I place an order for another jacket. But the made to order jacket was more expensive than a ready-made one, it may be over five thousand yuan. The made to order clothes have advantages in terms of fabric, style, color and size. I adopted his suggestion. So I had to order a tailored jacket. That was a most fashionable jacket.

The designer of the store took my measurements first, the measurements were coat length, chest, front width, back width, shoulder width, sleeve length, armhole, cuff, trousers length, waist, hip, thighs, knees bottom and so on. After taking my measurements, the designer told me he had conceived a new kind of collar, he asked me if I liked the new collar. I accepted his suggestion. The designer asked me to choose the materials. The store had an excellent stock of materials to choose from. They are woolen fabrics from all parts of the world. They are famous for their superior quality and patterns. I chose the one that was most suitable for me. He told me to come for the first fitting the next morning. Maybe he would make some alterations for the last fitting. He promised that he could supply my needs to the full.

New Words and Expressions

famous	著名的
suit	套装
be not available	没货
assistant	店员
sell out	售完
stock	备货
advantage	优势
inform sb.	通知某人
out of date	过时
suggest	建议
place an order for	订……货
front width	前胸宽
back width	后背宽
woolen fabric	羊毛织物
suitable	适合的
promise	承诺
to the full	完全地
armhole	袖底
accept	接受
shoulder width	肩宽
conceive	构思，设想
the made to order	定制的
superior	优良的
pattern	花型，图案
ready-made	现货
adopt	采纳
suggestion	建议
a pair of trousers	一条裤子
hip	臀围
thigh	横裆
knee	膝盖
have an excellent stock	备有优质的……
alteration	改动
supply	满足
in term of	在……方面
the first fitting	第一次试样

Notes to the Text 1

1. be not available 没有了

2. adopt one's suggestion 采纳某人的建议

3. have an excellent stock of materials to choose from 备有优质布料可供选择

4. I had to order a two-piece suit. 我得定制一套两件式套装（动词不定式作宾语）。

5. ...the designer asked me to choose... 设计师让我选择（动词不定式作直接宾语）

6. make some alterations ... 对……做些改动

7. The jacket was currently out of stock. 这夹克目前无货。

8. He would inform me once it was available. 一旦有货他会通知我。

Text 2

Placing an Order for a New Skirt 定做新裙子

Yesterday it occurred to me that I should buy a new skirt for myself. I went to a famous garment store nearby to see if there were any fashionable skirts for me. But the skirt I wanted was not available. It was sold out. The assistant told me that it was out of stock. The shop assistant recommend me another kind of skirt on display in the shop window. The skirt was the latest style. It was designed by a famous designer. It was popular with the customers with its special flavor. I don't think the skirt is my favor. I declined with thanks. He regretted not being able to supply me from stock. But there were a large stock of fashionable skirts and dresses for my choice. I looked at all the skirts and dresses, but I liked none of them. He suggested that I place an order for the skirt. Although the made to order skirt was more expensive than the ready-made one and the time was a little longer. I accepted his suggestion and ordered a skirt. There were a large choice of designed skirts. I chose a fashionable one among the skirts and dresses. The skirt was most attractive and very popular with young girls. I like the skirt.

The designer of the store led me to her work office and took my measurements. I looked at the pictures and patterns she drew. I pointed out some places I didn't like. I asked her to make some alterations. She did what I asked. She let me look at some fabrics and materials and asked me to choose the fabrics and materials. The store had an excellent stock of fabrics and materials to choose from. I chose silk for my skirt. And I chose some other materials. She told me to come for the fitting the next morning. Then she would give me a surprise. She said that she would make me an eye-catching skirt. She promised that the skirt would be the most beautiful.

New Words and Expressions

it occurred to sb. that	某人忽然想起	place an order for	订……的货
nearby	附近	a large choice	很多选择
be sold out	售完	point out	指出
recommend	推荐	have an excellent stock of	备有大量精美的……
out of stock	没货	ready-made	现成的
special	特别的	adopt	采纳
flavor	口味	promise	承诺
decline with thanks	谢绝	give sb. a surprise	给某人一个惊喜
regret	后悔	eye-catching	让人眼睛一亮
supply	提供	promise	承诺
lead to	领到……		

Notes to the Text 2

1. ... it occurred to me that... 我忽然想起……

2. I liked none of them. 我一件也不喜欢。

3. He suggested that I place an order...advice, suggest, insist, order, necessary, important, strange等词后面从句谓语动词用原形或should+动词原形。例如：We suggested that he study French. We insisted that he wear a white short. It is important that she keep her promise.

4. They regretted not being able to supply me from stock. 没有现货可供，他们感到非常抱歉。

5. The store had an excellent stock of fabrics and materials to choose from. 商店里备有大量精美的衣料和辅料可供选择。

6. She said that she would make me an eye-catching skirt. 她说她会给我做一件让人眼睛一亮的裙子。

Substitution Drills 替换练习

1. Yesterday I went to a garment store to order [a suit / a coat / a jacket].

2. [The colour / The style / The coat] I liked was not available.

3. He placed an order for [a coat / a jacket / a suit].

4. The store has an excellent stock of [materials / fashions / dresses] to choose from.

5. I will make some alterations at the [waist / chest / bottom / shoulders].

6. The [hip / chest / waist] was a little tight.

>> Exercises to the Texts 课文练习 <<

1. 抄写熟记本课单词和词组
2. 翻译下列短语

1）three-piece suit
2）first fitting
3）try on
4）stock of material
5）out of stock
6）adopt one's suggestion
7）be not available
8）inform sb.
9）ready-made
10）front width
11）make a few alterations
12）place an order for
13）the made to order
14）choose the material
15）last fitting
16）accept one's suggestion
17）eye-catching
18）like none of them
19）regret doing sth.
20）lead to

3. 将下列短语译成英语

1）现成的服装
2）备有优质衣料
3）最后试样
4）标出纽位
5）优质衣料
6）给某人量体
7）时间稍长点
8）指出
9）三条裤子
10）售完
11）定做新衣
12）做些改动
13）采纳了设计师的建议
14）过时
15）供你选择
16）选料
17）受……欢迎
18）吸引人的裙子
19）标出……的位置
20）底边

4. 翻译

1）昨天我去服装店定做了一套西服。
2）他在腰部做了些改动。
3）她没有采纳设计师的建议。
4）我喜欢的款式没有了。
5）设计师请我去选料。
6）定做的衣服比现成的贵得多。
7）我为儿子买了条裤子。
8）他们选了最好的布料。
9）我想定做一件大衣。
10）我们采纳了他们的意见。
11）这衬衫目前缺货。
12）我建议他学服装设计。
13）设计师为他量了体。
14）这些服装很受年轻人欢迎。

15）请标出纽扣的位置。

16）她采纳了设计师的建议。

17）我喜欢的裙子无货。

18）她建议小李学法语。

19）老师教学生们量体。

Conversation

>> Taking the Measurements 量体

A: I am keenly interested in measurement taking. It is said that you have rich working experience in this field. Could you tell me something about it? I want to learn something about it.

B: OK. I have worked as a designer in a clothing company. I think I am good at measurement taking. We must conform to local customs. We should respect the custom of the customers. Measurements are taken in the following ways: We take bust measurement round the fullest part of the breast with two fingers inside the tape measure. Keep the tape measure high up under the arms at the back.

A: How to take front width?

B: Front width should be taken across the front from where the sleeves are set in and at a level of 10 cm below the nape of the neck.

A: And how to take front length?

B: Front length is taken from the neck end of the shoulder to the center front.

A: What is waist measurement taken?

B: Waist measurement must be taken round the natural waist with a finger inside the tape measure.

A: Last question, could you tell me how to take sleeve length?

B: We take sleeve length from the top of the shoulder, over the back elbow and up the waist. We may have some other measurements taken in clothes, such as hip, across back, armhole width, armhole girth, wrist and so on.

A: Many thanks to you for what you have told me.

B: You are welcome. If you have any questions please ask me again.

New Words and Expressions

take the measurement	量尺寸
be keenly interested in...	特别感兴趣
measurement taking	量体（名词词组）
conform to	符合，顺应
local	当地的
it is said	据说
work as	当
bust	胸
full part of	最丰满处
breast	胸部
with two fingers inside	两手指放里面
keep...high up	望高持平
front width	胸宽
girth	周长
across back	背宽
sleeve	袖子
respect	尊重
set in	装（袖子）
level	水平
nape of the neck	颈根部
front length	前衣长
neck end of the shoulder	肩颈部
center front	前中
waist	腰
natural waist	腰最细处
sleeve length	袖长
top of the shoulder	袖山头
elbow	肘部
wrist	手腕
hip	臀围
What you said	你所说的

Reading Practice ❶

Ordering Some New Clothes
定做几件新衣服

Next month is Lucy' birthday. She is going to hold an evening party to celebrate her birthday. She wants to invite many her friends and classmates to attend the party. She wants to make herself beautiful and charming. She is going to order some new clothes this afternoon, including a full dress, a blouse, a shirt, a dress and a coat. She is going to a famous clothes shop. There are many experienced dressmakers and designers in the shop. They will give her some suggestions and advice. There are many excellent stocks of materials there, some are imported from Europe. She doesn't want to buy the ready clothes. She wants to order herself some beautiful and fashionable clothes.

The shop claimed that the prices of all the clothes were reasonable and the styles of the clothes were the most popular. At the garment store she can choose pieces of cloth for her dresses and blouse. The dressmakers there will match her dress and blouse perfectly. She can also ask a designer to take her measurements and design a new style for her. All the clothes she orders may expensive, but she can afford it, she doesn't care much about it. All that she wants is the best and fashionable.

New Words and Expressions

hold	举办	full dress	礼服
attend	参加	match	搭配
afford	花费得起	perfectly	完美地
care	在乎	claim	声称
reasonable	合理的	make oneself beautiful	把自己打扮漂亮

Reading Practice 2

Ordering a New Jacket
定做一件新夹克

Next week is Tom's birthday. He is planning to hold an evening party to celebrate his birthday. He will invite many his classmates and friends to attend his party. He wants to make himself handsome at the party. All his jackets are out of fashion. He doesn't want to buy a ready-made jacket. He wants to order a jacket with personality characteristics. He is sure everyone will be surprised when he puts on the jacket at the party.

He is going to a famous clothing store to order a new jacket. To meet the customers' needs, the shop offers a wide range of products with reasonable prices. A wide range of color, styles and patterns of clothes are available.There are many skilled and experienced tailors and designers in the store. They will give him some suggestions and advice. There are many excellent stocks of fabrics and materials, some of them are imported from foreign countries.

At the store he can choose something best for himself. The designer will recommend him a lot of styles and fabrics. The designer will takes his measurements and design several new styles for his choice. He will choose a style that is most suitable for him and order a jacket with personality characteristics. The jacket will put him on a great shine at the birthday party.

New Words and Expressions

hole an evening party	举办晚宴
celebrate	庆祝
handsome	英俊的
personality characteristics	个性特点
to meet the customers' needs	为了满足顾客的需求
offer	提供
a wide range of	广泛的
be available	有
recommend	推荐
put him on a great shine	使他大放异彩

Lesson Eleven

Text 1

The Assembling Procedure of a Blouse 女衬衣的缝制流程

A blouse is an out garment from neck to waist, usually with sleeves. It is worn by girls or women. It is kept in place at the waist with a belt or band. Before sewing a blouse, we should know the whole work. It should be in proper order, that is we should know which job must be done first, and then do the next. We should master the specific construction details of the blouse, such as its collar, darts, pleats, zips, pockets, tacks and so on. It will help us to get an overall picture of the order to assemble a dress, a shirt or a blouse, a skirt and a pair of trousers with these details. Besides, every worker on the assembly line must work hard because a single worker can hold up an entire assembly line if he doesn't keep up with the work flow.

The assembling procedure is always a logical one; that is, every step must be in order, take the body section of a blouse for example, the body is fitted and sewn before the sleeve is attached, otherwise the armhole in which the sleeve is inserted might not have the proper dimensions. Similarly, buttonholes and buttons are marked after the garment is completed, so that closures will be smooth and neat.

The assembling procedure of a blouse is as follows:

A. Body Section

1. Baste the darts.
2. Baste the seams.
3. Try on the blouse and adjust the darts and seams if it is necessary.
4. Stitch the darts and seams and press them.

B. Facings and collar

5. Attach the facing and fold back the self-facing at the centre front opening and press it.
6. Attach the collar to the neckline of the blouse and press it.

C. Sleeves

7. Attach the sleeves to the armholes of the blouse.
8. Attach the cuffs to the sleeves and press them.

D. Finishing touches

9. Mark the buttonholes and try on the blouse for positioning.
10. Baste the finished breast pocket to the blouse and check its position.
11. Try on the blouse for length and hem as well as press.
12. Sew the buttons on the blouse and cuffs.

New Words and Expressions

construction	结构
belt	皮带
band	布带
in proper order	按顺序
hold up	堵塞
work flow	流水线速度
that is	即，也就是说
detail	细节
dart	省道
zip	拉链
pleat	褶
overall	全面
assemble	装配
complete	完成
closure	关闭
keep up with	跟上
smooth	平整
neat	整洁
baste	假缝
seam	拼缝
adjust	调整
stitch	缝缉
press	熨烫
facing	过面，门襟
fold	折叠
self-facing	连门襟

procedure	步骤	center	中心
logical	逻辑的	opening	开襟
section	部分	neckline	领围
attach	绱，装	finish touch	收尾工序
insert	插进	breast	胸围
proper	恰当的	hem	底边
similarly	类似地	outer garment	外衣
position	位置	be kept in	在……束起
mark	标出	neck	颈部

Notes to the Text 1

1. It is kept in place at the waist with a belt or band. 腰部用皮带或布带束起。

2. We should master the specific construction details of the garments. 我们应掌握服装结构的细节。

3. It will help to get an overall picture of the order to assemble a dress. 这将有助于得到缝纫装配一件连衣裙的顺序。

4. A single worker can hold up an entire assembly line if he doesn't keep up with the work flow. 如果有一位工人跟不上流水线的速度就会拖住整条流水线。

5. The assembling procedure is always a logical one. 缝纫装配顺序始终按照一定的规律。

6. ...be in order 按照顺序

7. ...the body is fitted and sewn before the sleeve is attached. 大身部位合体并缝好后再装袖子。

8. Sew the buttons on the blouse and cuffs. 在衬衣和袖头处钉上纽扣。

Text 2

>> The Assembling Procedure of a Dress 连衣裙的缝制流程

A dress is an out garment. It is usually worn by girls or women. A worker who makes women's dresses is a dressmaker. The Assembling procedure of a dress is a logical one. First you should know how many procedures to make up a dress and how to do every procedure. You should know how many sewing steps and the quality requirements of every step. It is very important to know and understand all of these requirements. If you don't know what is the first step you shouldn't begin to do the sewing. You would get a failure result or make up the garment incorrectly. Quality is the first when we are working. In a word, we can't neglect the assembling procedure when we are working. The following are procedure for making up a dress:

1. Make the pockets, attach the pockets.
2. Make up the collar, cuffs (if there are any) and belt.
3. Attach the facings to the front pieces.
4. Stitch all the darts and the seams.
5. Attach the collar to the bodice. This is more easily done before the side seams of the bodice are stitched up.
6. Finish the ends of the sleeves and make waist openings, hems or attach cuffs.
7. Insert the sleeves.
8. Join the bodice to the skirt.
9. Work the side opening.
10. Fit on. Level and finish the hem.
11. Give a final pressing.
12. Sew on buttons, hooks and eyes.

New Words and Expressions

logical	合理的	join...to...	将……装到……
quality requirement	质量要求	hem	底边
in a word	总而言之	insert sleeve	装袖
understand	了解	work...	缝上……
neglect	忽视	fit on	试穿
failure result	不利的结果	level	使成水平
attach pocket	装配衣袋	final pressing	最后整件衣服熨烫
belt	腰带	finish the ends of the sleeve	完成袖子的所有工序
facing	门襟		

bodice	上身部	hooks and eyes	钩扣和扣眼
side seam	侧缝	ends of sleeve	袖口处
stitch up	缝合	waist opening	腰衩

Notes to the Text 2

1. You should know how many sewing steps and the quality requirements of every step. It is very important to know and understand all of these requirements. 你应该知道有多少步骤以及每一步的质量要求。了解掌握所有的这些需求是非常重要的。

2. You would get a failure result or make up the garment incorrectly. 你可能会得到失败的效果或做出了件不合格的衣服。

3. This is more easily done before the side seams of the bodice are stitched up. 衣身的侧缝缝合前这一步做起来更容易。

Substitution Drills 替换练习

1. Attach the [sleeve / collar / facing] to the [armhole / neckline / center front opening].

2. Baste the [darts / seams / breast pocket].

3. Mark the buttonholes on the [blouse / shirt / trousers / coat].

4. Stitch the [darts / seams / pockets / collar].

5. What size shoes / jacket / dress do you wear?

6. Try on the blouse / coat / dress for length.

Exercises to the Texts 课文练习

1. 抄写熟记本课单词和词组
2. 翻译下列短语
   1) the orders of sewing a garment
   2) to assemble a dress
   3) the body section of a dress
   4) fold back
   5) assembling procedure
   6) proper order
   7) darts and pleats
   8) hold up
   9) keep up with
   10) baste the seams
   11) self-facing
   12) front opening
   13) stitch the darts
   14) attach a sleeve
   15) try on
   16) attach a collar
   17) the armhole of the blouse
   18) mark the buttonholes
   19) press the sleeves
   20) breast pocket
3. 将下列短语译成英语
   1) 标出……位置
   2) 烫省道
   3) 收尾工程
   4) 装袖
   5) 缝纫步骤
   6) 做领
   7) 装袖头
   8) 腰衩
   9) 上衣领圈
   10) 连门襟
   11) 核对位置
   12) 大身部位
   13) 装领
   14) 朝后折
   15) 跟上
   16) 按照顺序
   17) 领围
   18) 缝口袋
   19) 标出胸袋
   20) 流水线速度

4. 翻译

1）我们应掌握服装结构的细节。
2）将领子装在大衣的领圈上，烫平。
3）请在这里标出省道的位置。
4）你穿多大尺码的夹克?
5）上衣做好后再钉扣。
6）这有助于得到全面信息。
7）请将袖克夫装到袖口上。
8）缝制服装应按照正确的顺序。
9）我们应该掌握服装设计的细节。
10）做袋然后装袋。
11）把衣领装到领圈上。
12）缝纫省道并烫平。
13）你穿多大尺码的衬衫?
14）试穿一下这件夹克看看长短。
15）我们必须了解夹克的装配顺序。
16）我们要跟上流水线的速度。
17）把纽扣钉到袖头上。
18）缝制所有的缝份。
19）设计师推荐他买那件上衣。
20）他以前当过设计师。

Conversation

>> Ordering a Blouse 定做女衬衫

A: Good morning, Madam Wang. What can I do for you?

B: I'd like to order a blouse, but little do I know about the popular fashions. Would you recommend me some styles?

A: How do you like this one? It's the latest style.

B: Oh, not so bad, but could you show me some other styles?

A: How about this one? I think it fits you well.

B: Ok. May I try it on?

A: Of course. This way please. Here is the fitting room.

B: I like the style, but it's a bit larger for me.

A: Let me see. It's too large round the waist. Here we should make a few alterations. The sleeves are a little shorter. This is not what you want. I'll take your measurements.

B: Thank you. It's kind of you to take my measurements. You know I pay great attention to my clothes.

A: Length is 70 cm, bust 99 cm, shoulder 41 cm, collar 39 cm. That's all, thank you, and what material do you want for your blouse?

B: I'm not sure, but I want something good. Could you help me?

A: All right. I think this kind of silk may be good for you. I often hear some people talking about the silk cloth. Some people call it "king cloth" .

B: I'd like this one. OK. That's settled.

A: Could you call in tomorrow for a fitting?

B: I'd be glad to.

A: After fitting, you'll get your blouse next week.

B: Many thanks.

New Words and Expressions

little do I know	不知……有关的情况	fitting room	试衣室
popular fashion	流行时装	that's all	就这些
recommend	推荐	silk	丝
fit you well	非常适合你	king cloth	料子王
try it on	试穿	that's settled	决定了

Reading Practice 1

A Nice Shirt 一件漂亮的女衬衫

Mr Brown (B) and his wife Mary Brown (M) are going to buy some skirts and jackets. Miss Li, a shop assistant is waiting on them.

L: How do you do? What can I do for you?

M:I'd like to buy some skirts and my husband wants to buy a jacket. Would you please show us some of your best?

L: All right. Here you are. The skirts are the best in our shop. They are the latest styles. Everyone here seems very interested in them. They are popular here. Our shop has a large choice of jackets. I think this jacket exactly fits you.

B: What a nice skirt! How smart the jacket is! They are very pretty, aren't they?

M:Yes, they are. But I don't think the color of the skirt fits me. Miss, would you please get me a black one.

L: Of course. How does this look?

M:I like it very much. It matches my blouse perfectly. How much do they?

L: The jacket is 500 yuan and the skirt is 320 yuan. 820 yuan in all.

B: Here is the money.

L: Here is the receipt. Our garments have a year guarantee. I hope you will come again.

B: We will. Good-bye.

L: Good-bye.

>> New Words and Expressions <<

wait on	服务	in all	总共
exactly	完美地	guarantee	保修期
smart	漂亮的		

Reading Practice 2

>> A Nice Dress 一件漂亮的连衣裙

This is a new dress. It is my sister's new dress. She designs and stitches the dress herself. She prides herself on her ability to design and stitch it. She thinks that it is the fashion of the year. But I don't think so. It may be out of date. Now my sister is ironing and pressing the dress to make it smooth and neat. The dress is very suitable for my sister. She is a careful dresser. I'm greatly interested in clothing. I also care much about dresses. I have a lot of dresses. It's difficult to fit a dress on me because I am irregularly built. I want to buy another new dress. I want a bright-colored dress. I would like to use theses dresses to dress up more beautiful. The style must be the most fashionable. It's up to me to decide whether to follow the fashion or dress in my own style.

I plan to go to the garment store to order a latest fashion or style. I am going to communicate with the designer and do my best to express my idea clearly. I am sure the designers there will meet my needs and design a dress for me. They will take my measurements and make some helpful suggestions. They will meet my requirements. The designers will ask me to choose the materials and my favourite cloth. They will recommend me some fabrics and materials. I will choose a piece of my favourite fabric. The fabric is suited for making dress. The store has an excellent stock of materials to choose from. It is necessary that they ask me to come for the fitting. They will make some alterations for the dress. They will ask me to come for the last fitting. The dress fits me well.

New Words and Expressions

pride oneself on	以……自豪
ability	能力
out of date	不流行
irregularly built	与常人不一样的身材
It's up to sb. to do	某人说了算
It's difficult to fit...on	很难有合适某人的……
decide	决定
iron	熨烫
communicate with	与……交流
express...clearly	清楚地表达……
smooth	平整的
dresser	穿……服装的人
bright-colored	鲜亮的
dress up	打扮
plan to do	打算做

Lesson Twelve

Text ❶

>> A Talk About Garments
服装报告会

Ladies and gentlemen,

Today I'm honored to be here to talk with you about clothing. I am very pleased to talk about garments. Garments are the things necessary for our daily life. Garments are things that people wear, such as coats, shirts, blouses, dresses, jackets, overcoats and so on. Most women like to have pretty dresses of the latest fashion and style, well-cut and stitched dresses, excellent underwear, and well-made shoes. That a young woman keeps up with current trends in clothes is necessary. Women, young ladies and girls are particular about their clothes. Now every store will display its unique style in order to attract the customers and compete in the market. In the morning people usually wear a blouse and skirt or a jumper and skirt. In the afternoon they generally wear a T-shirt and shorts or a dress. In spring and summer they like something lighter, and they wear a linen dress and a hat to match, wearing a pair of sunglasses. For sports they wear a short white linen dress, and for the sea side a beach dress. There are many bad customs and habits that ought to be abolished, such as wearing pajamas in the streets. It is not polite if one is not properly dressed in public. It is not polite wearing out of place in public. To some extent, garments reflect the inner quality and education of people.

Some rich people go to good tailors' for well-made clothes. There they can choose the patterns and the latest styles. And there is an excellent stock of materials to choose from. They can choose the one they want. The clothes designed and made up by the tailors and dressmakers are well-cut and keep their shape.

Most men go to department stores or supermarkets for ready-made clothes. The garments there are cheaper than tailors'. They can buy coats, jackets, vests, pants, shorts, ties, socks, trousers, waistcoat and pullovers. In winter they can buy overcoats, down wears, gloves and scarves, some people like tweeds or flannel garments. They are warm and prevent cold in winter. It's up to you to decide whether to follow the fashion or dress in your own style.

That's all. Thank you.

New Words and Expressions

be pleased to do 很高心做某事
be honored to do 很荣幸做某事
pretty 漂亮的
well-cut 裁剪好的
stitch 缝制
underwear 内衣
abolish 摒弃
keep up with 跟上
current 当时的
trends 潮流
be particular about 讲究
jumper 套衫
linen 亚麻
sunglasses 太阳镜
vest 背心
dress in 穿出
worn out 穿破
compete 竞争
to some extent 某种程度上
unique 独特的
beach 海滩
properly 恰当地
well-made 做工精致的
keep shape 不变形
pants 裤子
socks 短袜
waistcoat 马甲
pajamas 睡衣裤
pullover 套头毛衣
something lighter 薄的
down wear 羽绒服
scarf 围巾
tweeds and flannel 呢的和法兰绒的
out of place 穿着不整

Notes to the Text 1

1. ... display its unique style in order to attract the customers and compete in the market. 为了吸引顾客和在市场上竞争展示它的独特风格。

2. The clothes designed and made up by the tailors and dressmakers are well-cut and keep their shape. 师傅们设计和制作的男女服装裁剪得体不走样。

3. They are warm and prevent cold in winter. 这些服装冬天穿暖和防冷。

4. That a young woman keeps up with current trends in clothes is necessary. 年轻女士在服装跟上时代潮流是必要的。

5. It's up to you to decide whether to follow the fashion or dress in your own style. 是赶时髦还是穿出你自己的风格由你自己决定。

Text 2

>> A Talk About Fashion 时装报告会

Ladies and gentlemen,

I am very pleased to be asked to talk about fashion. Fashion is an interesting topic. The love of beauty is common to all people. Fashion is one of the most chief topics at the party. It gives young ladies and people their most interesting topic of conversation. When you are in bad mood, going to a supermarket to buy some clothes may be a good way to get rid of worries. Many people are in good mood after shopping. Clothes can reflect a person's character and education. To a degree, clothes reflects the inner quality of the people. Fashion is the popular style of clothes which is considered most to be admired and imitated during a period or at a place. People's clothes, interests and hobbies are usually relevant to social environment and their personal education background and experiences. Many people, especially young people and ladies are interested to talk about it. They often ask: "What kind of dresses are in fashion this year?" "What sort of hats are in fashion next year?"

Young people want to wear the latest fashion in the district. Businessmen try to produce more and more fashionable clothes to meet the young people's needs. In order to meet the needs of the customers, fashion designers do their best to design the latest fashions to meet their requirements. Many clothes out of fashion may come into fashion next year. Many clothes get out of fashion soon. We should not blindly follow the fashion trend. Everyone should have a unique style. We have our own styles. We should wear the garments which are most suitable for ourselves. I think the clothes one wears may stand for his social status. Everyone wear the clothes that can show his character, quality and spirits. Everybody will become the leader of fashion. Everybody can be a model. We have confidence in our ability. Come on!

Thank you.

New Words and Expressions

the love of beauty	对美的热爱	be in fashion	流行
common	共同的	district	地区
chief	主要的	out of fashion	过时
topic	话题	come into fashion	变流行
conversation	谈话	blindly	盲目地
be in bad mood	心情不好	follow	跟随
get rid of	去除	trend	潮流
character	性格	stand for	代表
to a degree	某种程度	status	地位
inner	内心	spirits	精神
quality	素质	leader	领导者
popular style	流行款式	model	模特
consider	认为	come on	加油
admire	欣赏	ability	才能
background	背景	get of fashion	不流行
imitate	模仿	environment	环境
a period	一段时间	a good way to do sth.	做某事的好方法

Notes to the Text 2

1. The love of beauty is common to all people. 爱美之心，人皆有之。

2. Clothes can reflect a person's character. To a degree, clothes reflect the inner quality of the people. 服装能反映一个人的性格。从某种程度上说，服装反映人的内心素质。

3. People's clothes, interests and hobbies are usually relevant to social environment and their personal education background and experiences. 人们的服饰、兴趣爱好往往与社会环境、教育背景和经历有关。

4. When you are in bad mood, going to a supermarket to buy some clothes may be a good way to get rid of worries. 当你心情不好的时候，去超市买几件衣服也许是去除烦恼的好方法。

Substitution Drills 替换练习

1. I am very pleased to talk about [clothes / garments / jackets / sewing machines].

2. Most ladies like to have pretty [dresses / skirts / overcoats] of the latest styles.

3. Most men go to tailor' for well cut [clothes / suits / trousers / jackets].

4. The [clothes / jackets / suits] designed and made up by the tailors are well-cut.

5. Many people go to the department store to buy [jackets / trousers / coats / shirts].

6. That suit looks [very good / too big / too small] on you.

Exercises to the Texts 课文练习

1. 抄写并熟记本课短语和词组
2. 翻译下列短语

1）jumper	11）matches well
2）well-made shoes	12）worn out
3）beach dress	13）prevent cold
4）latest fashion and style	14）flannel overcoat
5）out of place	15）to a degree
6）in order to	16）inner quality
7）look very good on sb.	17）out of fashion
8）keep shape	18）come on
9）keep up with	19）well-cut
10）be particular about	20）pretty skirt

3. 将下列短语译成英语

1）很高兴做某事	9）有信心
2）没现成的衣服	10）时装潮流
3）棉布花型	11）戴围巾
4）唯一的款式	12）内衣
5）有趣的话题	13）浅色的服装
6）被认为是	14）裁剪合体
7）欣赏和模仿	15）最新的款式
8）满足顾客的需求	16）吸引顾客

17）坏习惯

18）应该摒弃

19）穿着得体

20）穿睡衣

4. 翻译

1）玛丽去裁剪车间了。

2）他的上衣和裤子搭配得很好。

3）这位女装师制作的服装不走样。

4）大多数女士们喜欢最新款式的连衣裙。

5）为了吸引顾客他们设计了独特的款式。

6）男士们大多去超市买衣服。

7）你们班有多少人去过时装展览会。

8）时装展览会上展示了许多最新款式的大衣。

9）一个人的服饰与他的教育背景有关。

10）时装是个令人感兴趣的话题。

11）我们有信心学好英语。

12）我们不应该盲目跟随潮流。

13）商人们生产了很多服装以满足顾客的需求。

14）我们应该穿适合我们的衣服。

15）年轻人跟上时装潮流是必要的。

16）在公共场合穿着不得体是不礼貌的。

17）她对服装很讲究。

18）我很高兴来谈谈时装。

19）你穿这条裙子很漂亮。

20）这件夹克剪裁得很好。

Conversation

>> Introducing Our Company 介绍我们公司

A: Nice to meet you. I'm Wang Pin, the manger of the company.

B: Nice to meet you, too. I'm John Smith.

A: Welcome to our company. Our company is one of the largest companies in our city. It is divided into several sections, we have design research and developing department, men's and women's suits, out wear, children's wear and other important branches such as shirts, pajamas and sports wears, so the range of clothes is very diverse.

B: How many people are there in your company?

A: There are about four thousand people in our company, including workers, staff members, designers, engineers and technicians. They are trained personnel.

B: The output is very high every day, isn't it?

C: Yes, but we pay more attention to the quality and styles. We always keep an eye on the tendency of the fashion. Our designers make their efforts to design the latest styles to attract the customers. To meet the needs of the public, an order made department will be set up. We design and make up different kinds of clothes to meet the needs of different people. So the garments our company produced are popular with different people. Every day a great many coats, shirts, dresses and overcoats are made by the workers. Some of them are sold in clothes store, department stores and supermarkets. Some of them are exported to all parts of the world. We have won the great honor on the international market.

B: What you said impressed us deeply. Thank you for your introduction.

New Words and Expressions

out wear	外衣	be exported to	出口到
staff member	职员	trained	培训
be divided into	分为	technician	技术员
several sections	几个部门	output	产量
branch	分支	keep an eye on	盯着
range	范围	tendency	趋势
diverse	多样	set up	建立
personnel	人员	win great honor	赢得荣誉

Reading Practice 1

A Fashion Beginner 一位时装初学者

Li Tong worked in a garment factory of a big city. He has been educated in a garment school. He showed great interest in studying the art of fashion. He was very keen on art. He wanted to become a fashion designer. He realized that nothing but efforts ould translate his dream into reality. He made every effort to learn all the subjects well. He is sure with his talent, he is bound to design something good so long as he will work hard. He worked in the daytime and went to a spare-time school to learn design in the evening. He was confident that he had the ability to learn it well. After school, he surfed online to get some useful information and sometimes he talked about the topics and ideas with his friends on line. He learned how to draw fashion pictures and patterns. He also learned how to take measurements, how to cut out, sew and press all kinds of clothes. He often showed his patterns to his teacher and asked some questions. The teacher looked at the pictures and patterns he drew and corrected some mistakes he made in the pictures and patterns and asked him to draw them again. The teacher was satisfied with what he did.

Sometimes they discussed something about design. The teacher told him something popular in the world. Li Tong learned a lot from the teacher. He was sure that he would become a fashion designer in the future. That he never shrinks before difficulties surprised everyone around him.

New Words and Expressions

be bound to	一定	on line	网上
spare-time school	业余学校	topic	话题
fashion picture	时装画	shrink	退缩
be satisfied with	对……感到满意	surprise	使……惊讶
surf	冲浪	translate	翻译，转变
nothing but	除了……没别的	correct	改正
so long as	只要		

Reading Practice 2

A Clothes Beginner 一位服装初学者

Tom worked in a clothing company. He is a new worker. He shows great interest in clothing. He has great talent for drawing all kinds of coats, jackets, shirts, dresses, blouses, and overcoats. He also shows his great talent for art and clothes. He shows his character in his drawings. He wants to become a fashion designer. He believes that he is bound to succeed. He knows that talent and hard work are all important to career success. He goes to a spare-time art school to learn art and clothing design in the evening. He listens to the teacher carefully in class and draws fashion pictures and patterns at home. He learns how to take measurements, how to cut out paper patterns and how to lay out the paper patterns onto the materials and cut out them. He believes that he has the talent to make it. With the help of his teacher and classmates he tries his best to sew and stitch all kinds of clothes. He learns to design all kinds of clothes and shows the pictures to his teachers and asks them some questions. The teachers were satisfied with what he has done. The young student often receives great praise from the teachers. He knows that to learn and master this course involves perseverance. He is sure that he will become a fashion designer in the near future.

New Words and Expressions

show one's talent for	显示某人的才能	succeed	成功
show one's character in	显示某人的性格	be satisfied with	满意
lay out paper patterns onto	排料	praise	赞扬
what he has done	他所做的	master	掌握
in the near future	在不久的将来	involves	包括
have the talent to make it.	有成功的天分	perseverance	坚持，毅力

Lesson Thirteen

Text 1

>> Buying a Briefcase 购买手提包

A: Hi, how are you today?

B: I'm very well, thank you. When are you leaving for Shanghai?

A: Most probably next week.

B: Have you packed your things up?

A: No, I'm packing my things these days. I packed my clothes into two suitcases. My two suitcases are full of clothes and other things. There is no space.

B: I guess you want to buy another suitcase.

A: Yes, you're right. The suitcase I bought last time is not big enough, I need a bigger one. I have a lot of luggage to take this time. Do you have any good suggestions?

B: This time I would advise you to buy a leather trunk. We have three sizes for your choice, 27 inches, 30 inches, and 33 inches.

A: Is this ox hide or pigskin? Is the trunk strong?

B: It's pigskin. This kind of trunk is very strong and durable. And its capacity is big. Besides, four wheels are attached to it. So it's very convenient for you to carry it.

A: Do you know how heavy the medium size is?

B: About 3 kilos.

A: Good, I think I'll take one.

B: I would also advise you to buy a snake-skin belt and a crocodile-skin handbag. They are new products.

A: That's a good idea. Such goods are much more expensive in my country.

B: Look, the skin's color is natural. Both the belt and the handbag look quite smart.

A: Well, now how about this ox side briefcase? You surely need one since you have a lot of chances to travel on business.

B: The ox side appears to be very good in quality, but I do not have much confidence in the numerical lock.

A: This quality of the lock is excellent, it's better than any one.

B: The briefcase seems worth buying. OK, I take it.

A: You are our regular customer. You know all the things in our store.

B: Yes, I should say the best thing in your store is your excellent service.

A: It's our duty to do a good service for our customers heart and soul.

B: Thank you for your advice and good service. Good-bye.

A: Good-bye.

New Words and Expressions

briefcase	手提包	be attached to	附，装
space	空间	convenient	方便的
leave for	动身前往	medium size	中号尺寸
probably	可能地	crocodile-skin	鳄鱼皮
pack up	收拾行李	surely	一定
luggage	行李	have a chance to do sth.	有机会做某事
suitcase	衣箱	natural	自然的
guess	猜	appear	看上去
advise	建议	regular customer	常客，老客户
leather trunk	皮箱	do a good service for	为……提供优质服务
ox side	牛皮		
pigskin	猪皮	on business	出差
durable	耐用的	be full of	满了

besides	此外	service	服务
wheel	轮	heart and soul	全心全意
take	买了	confidence	自信
numerical lock	数码锁	capacity	容量
seem (be) worth doing	似乎值得买		

Notes to the Text 1

1. leave for A 动身去A地
2. leave A for B 离开A地去B地
3. Have you packed your things up? 你行李收拾好了吗?
4. You surely need one since you have a lot of chances to travel on business. 既然你有很多出差的机会那你一定需要一个。
5. I take it. 我买了。
6. It's our duty to do a good service for our customers heart and soul. 全心全意为顾客服务是我们的责任。

Text 2

Something About Blouses
有关女衬衫的对话

A: Well, Mr. Li, yesterday I went to the Clothing Fair. I saw the blouses and skirts made in your factory at the garment show, and I like them very much. Could you tell me something about them? I'm going to place an order for some with you.

B: Sure. We produce about 600,000 blouses and skirts each year. Over half of them are sold abroad and enjoy good fame there for novel designs, beautiful styles,high quality and reasonable prices. There are about twenty styles of them. They are made of silk, cotton and synthetic fabrics. We also have many blouses and skirts with traditional Chinese elements. These blouses and skirts are popular with young women. They are priced from three hundred yuan to eight hundred yuan.

That depends on its style, material and quality.

A: The price sounds reasonable. But would you permit me to have a look at the equipment and see how the blouses and skirts are made? I am interested in it. If your production permits, we are prepared to place larger orders with you.

B: You are welcome. Let me show you around our factory and workshop. This way, please. Here we are. Mr. Brown. This is our factory. We have more than 900 workers and staff members. The assembly lines are introduced from foreign country. We have computer aided design. The workers and staff members are all trained hard and strictly. Everyone pays close attention to the work flow. A single worker can hold up an entire assembly line if he does not keep up with the work flow. We pay much attention to the quality of products. The blouses and skirts we make are best in China.

A: OK, I feel great satisfaction in your production. Now let's look at your other workshops. I want to be further familiar with your factory and the supply position. I want to maintain a long-term business relationship with your factory.

New Words and Expressions

place an order for	订购	permit	允许
be sold abroad	销往国外	strictly	严格地
enjoy good fame	享有良好的声誉	production capacity	生产能力
novel design	新颖的设计	place larger orders with you	向你方大量订
reasonable	合理的	attention	注意
synthetic fabric	合成纤维织物	supply position	供应状况
with Chinese elements	有中国元素	maintain	保持
be priced	定价	long-term	长期
depend on	取决于	business relationship	贸易关系
aid	帮助	single	单个的
work flow	流水线流程	hold up	阻止
satisfaction	满意	entire	整个

Notes to the Text 2

1. Over half of them are sold abroad and enjoy good fame there for novel designs, beautiful styles,high quality and reasonable prices. 它们中一半以上销往国外，并以

新颖的设计，漂亮的款式，一流的质量，合理的价格在海外市场享有良好的声誉。

2. They are priced from three hundred yuan to eight hundred yuan. 它们的价格从300～800元不等。

3. If your production permits, we are prepared to place larger orders with you. 如果你们有生产能力，我们准备大量订货。

4. I want to be further familiar with your factory and the supply position. 我想对你们的厂和你们的供货状况作进一步了解。

Substitution Drills 替换练习

1. When are you leaving for [Shanghai / Beijing / Paris / London]?

2. I'm packing my [things / clothes] these days.

3. The [suit-case / overcoat / dress / suit] I bought last time is not big enough.

4. This time I would advice you to buy a [leather trunk / an overcoat / a blouse / a pair of trousers].

5. This dress is made of [silk / wool / cotton / nylon], isn't it?

6. Will it be [convenient / possible / important / necessary] for you to explain your plans?

Exercises to the Texts 课文练习

1. 抄写并熟记本课短语和词组
2. 翻译下列短语
   1) pack up
   2) be worth doing
   3) have a chance to do sth.
   4) appear to be
   5) be convenient for sb.
   6) be made of silk
   7) advice sb. to do sth.
   8) leave for
   9) leather trunk
   10) medium size
   11) travel abroad on business
   12) ox hide
   13) it is convenient to do
   14) heart and soul
   15) be priced
   16) long-term
   17) leave A for B
   18) novel design
   19) production capacity
   20) supply position
3. 将下列短语译成英语
   1) 动身去杭州
   2) 离开上海
   3) 猪皮夹克
   4) 优质服务
   5) 服装展销会
   6) 销往国外
   7) 享有良好的声誉
   8) 听起来合理
   9) 取决于材料
   10) 允许某人做某事
   11) 真丝的
   12) 中号
   13) 做某事是我们的责任
   14) 附带
   15) 耐用
   16) 出差旅行
   17) 贵得多
   18) 满了
   19) 值得去
   20) 两件行李
4. 翻译
   1) 我们有信心学好设计。
   2) 他没有足够的信心裁剪夹克。
   3) 我劝你去买件真丝衬衫。
   4) 他去年买的裤子短了。

5）你一定需要一件大衣。
6）那里购物很方便。
7）他们有机会出国留学。
8）这款大衣的质量是一流的。
9）这本书值得看。
10）我建议你学习英语。
11）你去那里有必要吗?
12）这条裙子是真丝的。
13）下周她动身去北京。
14）我们的服装都是销往国外的。
15）它们的价格定在500～1000元。
16）她对我们的生产感到很满意。
17）为顾客服务是我们的责任。
18）我想出国留学。
19）这些工人和工作人员都是经过严格培训的。
20）学习时装设计有必要吗?

Conversation

>> Some Common Threads 常用缝纫线

A: It is said that there are many kinds of sewing threads, what are they? How many kinds of threads we have in sewing and stitching. Could you tell me something about the threads? I'm very interested in them.

B: Yes. Thread is a length of very thin made by spinning cotton, silk, man-made or synthetic fibers or wool, used in sewing. We usually use cotton thread, silk thread and synthetic thread. Cotton thread and silk thread are made of natural fibers. Synthetic threads are made of synthetic fibers. Every kind of thread has its own function and usage.

A: How to select sewing threads?

B: The selection of sewing threads depends on the material which is to assemble. Threads are made from a wide variety of materials, so every thread is appropriate to its material. Suitable thread gives a better look. People usually choose cotton thread to sew cotton cloth, silk thread for silk cloth, and synthetic thread for synthetic cloth.

A: If we use thread which is not suitable for material, what will happen?

B: If a garment is sewn with a thread which is not suitable for the fabric, something bad will happen. It may be unsatisfactory in wear, and may give a bad look. The bad results of garments will happen.

A: What are the advantages of synthetic threads?

B: Synthetic threads are widely used in sewing synthetic fabrics. They are strong, smooth and durable. Synthetic threads are also elastic.

A: What are the advantages of natural threads?

B: They are soft and warm but not strong as synthetic threads, so they are sometimes blended with synthetic fibers.

A: Now I understand that different fabrics need different threads in order to get a better result and satisfactory wear.

New Words and Expressions

sewing thread	缝纫线	a wide variety of	大量
spin	纺（纱）	it is said	据说
a length of	一长条	appropriate	合适的
synthetic	合成的	unsatisfactory	不满意
natural fiber	天然纤维	advantage	优点
select	选择（动词）	durable	耐用的
selection	选择（名词）	elastic	弹性
assemble	缝制	be blended with	与……混纺
appearance	外观	usage	用途
function	功能		

Reading Practice ❶

Visiting a Garment Factory 参观服装厂

Yesterday our teacher told us we were going to visit a garment factory. We had been looking forward to it for a long time. When our teacher told us that the head of the factory kindly permitted us to visit, we were very excited. The factory had been newly built. The equipment of the factory was mostly made by our country and some introduced from foreign country. They tried their best to realize the innovation in technology. They equipped their workers with the latest sewing machines and steam irons. Some of them are invented by themselves. Everyone was specially trained for his work. They had computer-aided design (CAD) and advanced operation process. After the new equipment and technique were introduced, the factory produced twice as many garments as the year before. Their products were often in short supply and out of stock.

The head of the factory showed us around the factory and replied to the questions we asked. When we entered the workshop, the workers were working hard by the assembliny lines. Everyone here was busy doing his own job. Some were making collars, some were stitching darts, yokes and pockets, while others were attaching collars and sleeves, their action was so skilled and perfect, the work flow was so fast. We were surprised to watch their movements and operation. We looked at their garments happily. We hope we would become a good worker soon. We were impressed by the workers' enthusiasm for labor.

New Words and Expressions

a wide variety of	大量	technique	技术
innovation	革新	reply to	回答
appropriate	合适的	assembling line	流水线
specially	特别地	dart	省道
computer-aided design	计算机辅助设计	yoke	育克，过肩
advanced	先进的	attach collar	装领子
operation process	作业流程	impress	留下印象
technology	技术	enthusiasm	热情

Reading Practice 2

A Garment Factory 服装厂

Last week our teachers took us to visit a garment factory. We had been looking forward to it for a long time. When our class teacher told us that the head of the factory had permitted us to visit the garment factory, we were very excited. The factory had been newly built last few years. The equipment of the factory was all introduced from foreign country. After the new equipment and technique were introduced, the factory produced twice as many garments as the other factory.

We were sent to the garment factory in a bus. When we went into the workshop, we were surprised to see there were so many workers sitting there working seriously and attentively. The cloth and pieces in their hands soon became garments. The head of the factory told us to obey the instructions and follow the rules of safety. We looked at carefully what the workers did.

The workers were working hard by the assembling line. Everyone was busy doing his own job. Some were making collars, some were stitching darts and cuffs. They each worked carefully and attentively. There were many fine and perfect garments on the work flow. We were very interested in the way they stitched and sewed. We began to realize that it was not easy to do every job well. We asked many questions about the operation and the products. The head of the factory replied the questions we asked and showed us the garments. The high quality of the products left a deep impression on us.

New Words and Expressions

the head of the factory	厂长	impression	印象
seriously	认真地	instruction	指示，指导
attentively	注意	begin to do	开始做
assembly line	流水线	realize	意识到

Lesson Fourteen

Text 1

>> Some Sewing Equipment 一些缝纫设备

A Chinese proverb goes "If a worker wants to do a good job, he must sharpen his tools first." It is necessary to have the right tools before making something well. So we should have right tools and equipment before stitching and sewing a coat, a dress, a jacket and so on. Generally speaking, some modern equipment is absolutely necessary, by holding modern advanced equipment and professional technicians we will produce top quality products. Sewing machines, scissors, needles, thimbles, tape-measures, and irons are essential. Besides there are rulers, tailor's chalks and so on. Sometimes we have to invent some useful and necessary machines, tools and equipment. We should buy the best that can be afforded, look after them well and they will last a long time. The useful and essential equipment needed will be as follows:

1. Sewing Machines

A sewing machine is an expensive piece of equipment. It is the most important equipment in our production. If we use and take care it properly, it should last a long time. A good sewing machine will help us a lot. We should oil it every day. There are some other sewing machines, such as lock stitch machines, zigzag machines, buttonhole machines, safety stitch over lock machines and so on.

2. Scissors

Scissors are common and useful tools in our production. A pair of cutting-out scissors is about 20 cm long or even longer. A smaller pair is for cutting cloth and buttonholes. A small pair with both ends pointed is for thread and embroidery. Never use scissors for any purpose other than that for which they are intended and keep them sharpened always.

3. Needles

There are many different needles in tailoring and dressmaking. We should have a good assorted supply of needles. Both sewing machine needles and hand needles are necessary. Needles are numbered for size, high number for fine ones and low number for coarser types. We should keep them off water.

4. Thimbles

Usually we have two kinds of thimbles, that is, white thimbles and silver ones. White metal thimble give the longest wear, silver ones are too soft and too easily pierced. A thimble should fit comfortably.

5. Tape Measure

Tape measures are absolutely necessary in measurement taking. Tape measure should be of good quality, otherwise it will stretch. If no hem-marker is available, a long ruler is useful for leveling hems.

6. Irons

Irons should be comfortable to handle and should weigh about seven pounds. The steam irons are better. An iron which can be regulated may be left with no fear of scorching.

>> New Words and Expressions <<

proverb	谚语
generally speaking	一般来说
absolutely	绝对地
invent	发明
essential	必要的
last	持续
as follows	如下
properly	恰当地
regulation	规则
sharpen	使变得锋利

lock	锁	keep off	远离
zigzag	之字形	silver	银的
both ends pointed	两头尖	pierce	刺穿
other than	除此之外	hold	持有，拥有
intend	打算	fit comfortably	用着舒适
keep sharpened	保持锋利	stretch	卷曲
assorted	各种各样的	hem-marker	底边标尺
supply	供应	level	使成水平
be numbered for size	用数字表明尺寸	regulate	调节
fine	细的	fear	害怕
technician	技术人员	scorch	烫焦
coarse	粗糙的	comfortably	舒适地

Notes to the Text 1

1. A Chinese proverb goes "If a worker wants to do a good job, he must sharpen his tools first." It is necessary to have the right tools before making something well. 有一句中国古语："工欲善其事，必先利其器。"要想把事情做好，有合适的工具是必要的。

2. ...oil it (sewing machine) every day. 每天给缝纫机加油。

3. ... buy the best that can be afforded, look after them well and they will last a long time. 购买你所能支付得起的最好工具，好好加以保养，这样你的工具将会经久耐用。

4. The essential equipment needed will be as follows. 所需工具如下。

5. Even longer 更长。even, much, still, far, a lot, no, a little, a bit 等修饰形容词或副词的比较级。

6. We should have a good assorted supply of needles. 我们应该备有各种各样的针。

7. Never use scissors for any purpose other than that for which they are intended and keep them sharpened always. 千万别将剪刀用于规定以外的任何用途，并经常保持锋利。

8. An iron which can be regulated may be left with no fear of scorching. 一种可以调节温度的熨斗不用担心烫焦衣物。

Text ❷

>> Sewing Machines and Scissors
缝纫机和剪刀

If we want to do a good job, it is important to have right tools and equipment. We need many tools and much equipment in production. Some companies and factories furnish everything we need for sewing and stitching. We can choose from them carefully. We have made a lot of rules and regulations for the correct use of tools and equipment. We should try our best to reduce unnecessary slips and mistakes. But some people think it is unnecessary to have the right tools before sewing and stitching a garment. They don't know the importance of the sewing equipment. Nothing is more important than to have right equipment before sewing and stitching garments. We should take care of these machines and equipment in our production. This is a matter of importance. Now let's look at some useful equipment.

1. Sewing Machines

Sewing machines are our most commonly used equipment. Much effort is needed to operate sewing machines well. We develop many rules and regulation in operation. These rules and regulations must be strictly observed. If it is given proper care and usage, it should last a long time. We should focus our attention on safety in production. We should take care of our sewing machines. If we take good care of our sewing machines, we'll make the machines run safely for a long time and we can increase our production. We should clean it and oil it every day.

2. Scissors

Scissors are the most commonly used tools. We use them to cut out paper patterns, to trim the pieces of the clothes, to cut out some small pieces and so on. Each pair of scissors has its own purpose. We must keep our scissors sharpened. We mustn't use our scissors for any other purpose than that for which they are intended.

New Words and Expressions

proper	恰当的	trim	修剪
furnish	供应	intend	打算
choose from	从中挑选	slip	失误，差错
daily	每日的	unnecessary	不必要的
focus on	集中	strictly	严格地

Notes to the Text 2

1. We have made a lot of rules and regulations for the correct use of tools and equipment. 我们对正确使用工具设备制定了很多规章制度。

2. Much effort is needed to operate sewing machines well. 操作好缝纫机需要很多努力。

3. We mustn't use our scissors for any other purpose than that for which they are intended. 我们不能用这些剪刀剪不该剪的东西。

Substitution Drills 替换练习

1. They had a talk about the clothes / jeans / evening dress / fashions .

2. I look forward to going abroad / visiting the fair / studying abroad .

3. It seems a shade tight round the waist / breast / hip .

4. Never use | scissors / sewing machine / iron | for any purpose.

5. | Tape measure / Scissors / Steam iron | should be of good quality.

Exercises to the Texts 课文练习

1. 抄写并熟记本课短语和词组
2. 翻译下列短语
   1）hem-marker
   2）other than
   3）regulated iron
   4）fit comfortably
   5）tailor's chalk
   6）last a long time
   7）as follows
   8）such as
   9）common and useful
   10）have a good assorted supply
   11）high number
   12）keep sth. sharpened
   13）can be afforded
   14）metal thimble
   15）tools and equipment
   16）make a lot of regulations
   17）operate sewing machines
   18）increase our production
   19）trim the pieces of the clothes
   20）keep the scissors sharpened
3. 将下列短语译成英语
   1）相配
   2）如下
   3）两头尖
   4）使锋利
   5）常用工具
   6）修剪
   7）不用担心烫焦衣物
   8）一般来说
   9）完全必要
   10）照看好
   11）更长
   12）不必担心
   13）恰当地
   14）两把剪刀
   15）给他很多帮助
   16）钉扣机
   17）类似的剪刀
   18）卷曲
   19）可调温度的熨斗
   20）必需的设备

4. 翻译

1）每天给缝纫机加油是必要的。

2）剪刀要保持锋利。

3）这种布料很容易伸长。

4）这条裤子和你的衬衣很相配。

5）我们有各种颜色和尺寸的衬衣。

6）熨斗保养得好能用很长时间。

7）新的缝纫机零件是蓝的。

8）所有的设备都是优质的。

9）剪刀的质量要好。

10）他们备有各种缝纫机。

11）我们应该购买最好的设备。

12）我们应保持剪刀锋利。

13）顶针应戴着舒服。

14）他们有许多可调节温度的熨斗。

15）我盼望着去国外留学。

16）缝纫机是我们生产中最重要的设备。

17）我们学校制定了新的规章制度。

18）我们不能用剪刀剪不该剪的东西。

19）臀部好像有点紧。

Conversation

>> The Importance of Temporary Stitches
临时针迹的重要性

A: In today's lesson our teacher told us when making up a garment in some areas of the garment we should use temporary stitches. Would you please tell me what the temporary stitches are and why we should use them when we are making up a garment?

B: Temporary stitches, as the name suggests, are stitches which last for a short time only in the production. When the garment is finished, the stitches will be removed. Temporary stitches are used to translate the fitting lines from patterns onto materials for tacking parts or pieces together before stitching. Sometimes the temporary stitches are used to hold the pieces or parts temporarily. It used as a guide for machining and for holding parts such as pleats temporarily in position. When the garment is finished, they are all removed.

A: Is it important to use it in tailoring and dressmaking?

B: Yes, although the stitches are temporary we can't ignore the importance of them. It contains some traditional Chinese elements. It shows Chinese culture and technique and skill. It is very important to learn this skill. Their importance cannot be over estimated and we should pay special attention to it.

A: Could you tell me how to do it?

B: Yes. Let me show you how to do it. Choose a needle which can be threaded easily and thread it with long double cotton thread. We should always use tacking cotton as it is fluffy and stays in the material as other threads do not. Take a smooth stitch through a hole in the pattern and through both layers of materials. Take a stitch back in the same hole and pull the cotton through. Pass on the next hole and repeat. Cut the threads between each hole.

A: I see. Thank you.

New Words and Expressions

temporary	临时的	remove	拆除
area	部位	estimate	评价
translate	表达	tailor tacking	粗缝
temporarily	临时地	thread easily	容易穿线
fitting line	净线，净缝	thread	穿针
tack...together	将……粗缝在一起	double cotton thread	双股棉线
parts and pieces	部件和衣片	fluffy	有毛的
pleats	褶裥	take a smooth stitch	挑缝一小针
hold...in position	钉在……位置上	layer	层
pull	拉动	take a stitch back	往后缝一针
stay	留在	pass on	向前
repeat	重复	ignore	忽视

Reading Practice ❶

A Welcome Speech 欢迎词

Ladies and gentlemen,

Now we are very happy to have Professor Brown with us and give us a wonderful lecture on how to prepare for design. Professor Brown is a specialist in clothes and a scholar of art. He used to be a member of the Clothing Association in Paris. Right now he is teaching in the Garment Institute in our city. He has been engaged in teaching clothing design for quite a few years and has much study of it. Apart from fashion design, he is also well versed in English and French. His education and eloquence as well as his kindness made his lecture so attractive and touching that everybody likes him very much. He encourages students to think independently. The students of his school are fascinated by his creative ideas. He designed many beautiful and smart clothes, some of them won the highest prize in International Garment Contest. Today we have the honour to ask him to give us a lecture.

Technique and knowledge are of great importance as far as the speeding up of our production is concerned. After the lecture Professor Brown will meet and talk to the students. We can ask him any questions about design and clothes. Professor Brown will answer your questions in detail. We should learn from his noble spirit and do our best to study hard for our motherland. I'm sure we will greatly benefit from his lecture.

Now, let's warmly welcome Professor Brown.

New Words and Expressions

specialist	专家	benefit	获益，获利
scholar	学者	have much study of	有很深的造诣
association	协会	eloquence	雄辩
institute	学院	attractive	吸引人的
be engaged in	从事于	noble	高贵的
apart from	除……外	be well versed in	精通
speed up	促进，加速	as far as...is concerned	就……而言
encourage	鼓励	independently	独立地
creative idea	创新思想	fascinate	吸引

Reading Practice 2

Making a Welcome Speech 致欢迎词

Ladies and gentlemen,

Now we are pleased to have Miss Smith with us and give us a wonderful lecture on how to prepare for design. Miss Smith is a famous costume designer. Her works are very popular with young people and students. Here is a wonderful piece of her work. She always encourages young people to have creative idea, to think independently. Her designs will be of interest to your students. The designs of her products are more interesting than other designers. I am sure her designs will appeal to customers. If her designs do not appeal to you, she will show you the others. Maybe she will design some clothes with you together on the spot.

Miss Smith has been engaged in teaching clothes design for a long time and has much study of it. She has made outstanding achievements with great courage,confidence and creativity in the past few years. Her education and eloquence as well as her sense of humor made her lecture so attractive and touching that everyone likes her very much.

After the lecture Miss Smith will meet and talk to the students. We can ask her any questions about art, drawing, design, measurement taking and clothes materials as well as the issue make-up. She will be happy to answer your questions. I am sure we will benefit from the lecture and talk.

Now, let's warmly welcome Miss Smith.

>> New Words and Expressions <<

costume 服装
appeal to 吸引
courage 勇气
outstanding 杰出
achievement 成就
creativity 创造力
on the spot 当场
issue 问题
make-up 化妆

Lesson Fifteen

Text ❶

>> Cotton Garments 棉布服装

Cotton garments are known for their soft, warm and comfortable. Nowadays cotton garments are more and more popular with people,and are getting more and more popular in many places, local area and other countries. Colorful and fashionable cotton garments add to a lively atmosphere in spring and summer streets. The cotton clothes make people feel warm and comfortable. Wearing the cotton clothes makes people feel relaxed. Recently more and more people have come to realize the comfort of cotton garments. People prefer wearing cotton garments, such as trousers, coats, jackets, and sports wear. Some people think wearing cotton garments is a social status. We have a lot of cotton garments such as jackets, trousers, shirts coats and underpants. Being one of the daily necessities in human life, cotton garments are soft, warm, plain and comfortable. Cotton garments are well received in every market. In many shops, the cotton garments sell well.

We export cotton garments to all the countries of the world. These clothes are made of exquisite materials, fine workmanship and first-class quality. They are well received by the customers. They are fine in texture and durable in use. They are made out of bleached, dyed, printed, jacquard or yarn-dyed poplin, jean, drill, khaki, etc. in various counts. Nowadays, wherever you go in the world, you can see clothes made in China, especially cotton garments.

Different varieties of cotton garments are designed to fit different needs. For instance, the denim trousers are reinforced with double lines at all right parts. The jackets are smart cut and comfort to wear, while the ski-suits and swimming are specially designed for their respective uses.

Best selected material, attractive design, first class workmanship and all of our products are of good quality — that is why Chinese cotton garments are so welcome everywhere.

New Words and Expressions

nowadays	现在	be well received	很受欢迎
recently	最近	necessity	必需品
lively	活泼的	atmosphere	气氛
colorful	各色的	add to	增添
realize	认识到	being	作为
come to	开始	drill	斜纹布
underpants	内裤	khaki	卡其布
plain	平整的	counts	支数
texture	纹理	denim	粗斜纹布
durable	耐用的	reinforced	加固的
bleached	漂白的	ski-suits	滑雪衣
dyed	被染色的	respective	各自的
printed	印花的	selected	挑选的
jacquard	提花织物	workmanship	工艺
yarn-dyed	条染	everywhere	到处
jean	细斜纹布	sell well	卖得好

Notes to the Text 1

1. Being one of the daily necessities in human life. 作为人们日常生活必需品之一。

2. Colorful and fashionable cotton garments add to a lively atmosphere in spring and summer streets. 多彩时髦的纯棉服装给春夏的街道增添了活泼气氛。

3. Nowadays, wherever you go in the world, you can see clothes made in China. 现

如今，无论你走在世界哪个地方都能看到中国服装。

4. ...are smart cut and comfort to wear. 裁剪得体，穿着舒适。

5. plain and comfortable 朴素而舒适

6. They are made out of bleached, dyed, printed, jacquard or yarn-dyed poplin, jean, drill, khaki, etc. in various counts. 它们由不同支数的漂白布、染色布、印花布、纱染织物、提花府绸及斜纹卡其等组成。

7. Different varieties of cotton garments are designed to fit different needs. 不同种类的棉布服装是根据不同的需要而设计的。

Text 2

Comfortable Cotton Garments 舒适的棉布服装

Many people and customers have recently expressed great interest in cotton garments. Cotton garments are soft, warm, natural and comfortable. Nowadays many people think that wearing cotton garments is a kind of fashion style. Cotton garment sometimes are regarded as a social status. Some people think cotton garments stand for a sense of satisfaction. Cotton clothes are well received in the market. They are soft and comfortable and are popular in the market. There are many cotton clothes such as jackets, coats, shirts, trousers and underpants. Cotton clothes have become one of the daily necessities in human life. Many companies and factories produce more and more cotton garments to meet the customer's needs.

We have exported a lot of cotton clothes and clothing to many countries all of the world. Our cotton products are well received in the world market. They are fine in texture and durable in use. We have bleached, dyed, printed, jean, drill, khaki, etc. in various counts.

We have different varieties of cotton garments to meet different needs, for example, we have denim trousers which are reinforced with double lines at all right parts; we have jackets which are smart cut and comfort to wear.

We have selected the best materials. We have attractive design and first class workmanship. That is why our cotton garments are well received everywhere.

New Words and Expressions

express	表达	texture	质地
natural	自然的	poplin	府绸
be regarded as	被认为	drill	厚棉织品
status	地位	export to	出口到
stand for	代表	various	各种
a sense of	感	variety	种种
nowadays	现今	double lines	双线
be well received	获得好评	right part	正面
soft	柔软	smart cut	裁剪合体
necessities	必需品	select	选
human	人类	exquisite	精美的，精致的

Notes to the Text 2

1. Many people and customers have recently expressed great interest in cotton garments. 最近很多人对棉布服装表现出很大兴趣。

2. Nowadays many people think that wearing cotton garments is a kind of fashion style. Cotton garment sometimes are regarded as a social status. 现今许多人认为穿棉布服装是一种时尚范。棉布服装有时被认为是社会地位的标志。

3. Some people think cotton garments stand for a sense of satisfaction. 有人认为棉布服装代表了一种满足感。

4. They are fine in texture and durable in use. 它们质地优良，经久耐用。

Substitution Drills 替换练习

1. There is a great variety of [cotton / woolen / synthetic] clothing such as jackets, trousers and dresses.

2. Cotton / Woolen clothing is warm, plain and comfortable.

3. The jackets / overcoats / suits are smart and comfort to wear.

4. There are all kinds of bleached / dyed / printed cotton fabrics.

5. Nowadays more and more people come to know the importance of learning English / computer / swimming well.

6. Look at the workmanship on this jacket / shirt / coat / overcoat.

Exercises to the Texts 课文练习

1. 抄写并熟记本课短语和词组
2. 翻译下列短语
   1）cotton clothing
   2）dyed poplin
   3）ski-suit
   4）import from
   5）sell well
   6）double lines
   7）be well received
   8）come to realize
   9）export to
   10）smart cut
   11）daily necessities
   12）attractive design
   13）fit different needs
   14）come to know
   15）be regarded as
   16）stand for
   17）comfort to wear
   18）first class workmanship
   19）durable in use
   20）various counts

3. 将下列短语译成英语

1）受到……欢迎
2）开始认识到
3）日常必需品
4）质地优良
5）经久耐用
6）印花棉布
7）提花织物
8）粗斜纹布
9）条染的
10）精心挑选
11）双线加固
12）为……而特别设计
13）对……表示有兴趣
14）棉布产品
15）满足不同的需要
16）代表
17）卡其
18）世界市场
19）滑雪衣
20）各种支数

4. 翻译

1）不同的服装根据不同的需要而设计。
2）那里陈列着不同支数的棉布。
3）我们每年出口大量服装到世界各国。
4）这些布料质地精良，结实耐用。
5）用双线在裤子后缝加固。
6）他买的这件衣服朴实舒适。
7）这条裤子的正面都用双线加固。
8）我们的服装工艺是一流的。
9）棉布服装以柔软保暖舒适而著称。
10）我们每年向那个国家出口棉布夹克。
11）人们更愿意穿纯棉服装。
12）这些上衣受到顾客们的欢迎。
13）这些服装裁剪得体，穿着舒适。
14）这些夹克朴素而舒适。
15）她看了看裙子的做工工艺。
16）这些衣服都是生活必需品。
17）学生们开始认识到学好外语的重要性。
18）请在正面加固。
19）现如今，无论你走在世界哪个地方都能看到中国服装。
20）他们的服装有着吸引人的设计。

Conversation

>> A Dialogue About Interlining 关于衬的对话

A: When making up a garment we often put some interlinings on some special areas, what's an interlining? Why do we need interlinings when making up garments?

B: An interlining is an additional piece of fabric which is applied to particular garment sections or areas to support the garments and garment sections or areas.

A: Is it necessary for us to apply interlining to garment?

B: Yes, applying interlining to garment is necessary and very important. In some garment sections and areas interlining is indispensable. We apply interlining to garments in order to require further support, stabilize garment sections or areas, reinforce sections, give shape to collars and lapels and prevent seams impressions.

A: How many kinds of interlinings are there do you know?

B: There are many kinds of interlinings in our tailoring and dressmaking.They can be divided into three kinds of interlinings according to the weaving method, woven interlining, non-woven interlining and knitted interlining. Non-woven interlining is also called adhesive interlining or non-woven adhesive interlining.

A: What are the advantages of these interlinings?

B: The main advantage of woven interlinings is that they can be cut exactly the same grain as the pieces to which they will be applied. The advantages of non-woven interlinings are cheap, and they can be cut multi-directionally. Knitted interlinings can be stretched as well as controlled.

A: Are they fusible?

B: Woven interlinings are not fusible, non-woven interlinings are fusible and knitted interlinings are not fusible. Fusible interlinings are widely used in garment industry. The application of fusible interlinings makes us produce excellent clothes.

New Words and Expressions

adhesive	黏合的
interlining	衬头
additional	额外
lapel	驳头
impression	痕迹
woven	织造
apply to	应用
particular	特别的
require	获得
indispensable	必不可少的
further support	进一步支撑
stabilize	稳定
reinforce	加固
give shape to	使定型
non-woven	无纺
knitted	针织
multi	多方
stretch	延长
fusible	热熔的
application	应用
directional	方向的
control	控制

Reading Practice ❶

Our Company 我们的公司

Today we are going to tell you something about our garment company. Our company is one of the largest companies in our country. It is divided into several sections. Our company mainly produces men's and women's suit, light clothing, outwear and other clothes such as shirts, pajamas and sports wears, so that the range of merchandise is very diverse. We also handle both the import and export of clothes. We have established trade relations with the firms of over one hundred countries in the world. As our export garments are all manufactured in our own factories, we have advantages in prices. Our products are enjoying fast sales on the market. Our company enjoys a very high international repute.

There are about nine thousand people in our garment company, including designers, engineers, technicians, staff members and workers. Every day a great many coats, shirts, dresses and overcoats are made by the workers. Some of them are sold to garment stores and department stores, and some of them are exported to all parts of the world. We introduced computers to our design. Computers have brought about many changes in our company. They have been of great use in design and production. We are proud of the modern machines that were being used.

Our company is also an international trade company. I don't think that the goods of other company's can compare with ours. Our products can compete with any other similar products.

New Words and Expressions

be divided into	分为	handle	从事
outwear	外衣	establish	建立
pajamas	睡衣	compare with	与……相比较
range	范围	firm	公司
enjoy fast sales	很畅销	manufacture	制造
merchandise	商品	technician	技术员
diverse	多种多样	export	出口

import	进口	produce	生产（动词）
bring about	带来	enjoy a high repute	享有很高的声誉
production	生产（名词）	compete with	与……相竞争

Reading Practice ❷

>> Introducing Our Company 介绍我们公司

Let me tell you something about our company. Our company was founded more than 50 years ago. Starting from a small factory, our company has become a modern enterprise with complete equipment and complete products due to our constant absorption of all kinds of talents, constant innovation and invention. We have intellectual property rights and patented inventions. Now our company is a large international trade company. Our company is specialized in handling the import and export business in sewing machines, clothes, steam irons and other pressing equipment.

In the past few years our company has invented many kinds of sewing machines, equipment and other machines. We are most grateful to our engineers, staff members and workers and cadres. Our company has managed to recruit all kinds of talents. Emphasis on talents, education and the quality of products made our company stronger and stronger.We have been handling all kinds of fabrics and garment materials for more than twenty years. We are engaged in the exportation of clothing materials and textiles. Our exports are well received in the world market. We have established trade relations with the firms and companies of over one hundred countries in the world. As our exports are all manufactured in our own factories, we have advantages in price. All the products are of good quality. Everyone knows that the price depends on the quality, but our prices are reasonable. Our products enjoy great popularity in the world and a very high international repute. So our products sell well in the market.

New Words and Expressions

trade	贸易	patented	专利的
emphasis on	强调	due to	由于
enterprise	企业	found	成立
talent	人才	be grateful to	感谢
complete	完全的	recruit	招募
constant	不断的	import	进口
absorption	吸收	be specialized in	专门
innovation	革新	be well received	很受欢迎
intellectual	知识的	repute	名誉，名声
property	著作权		

Lesson Sixteen

Text 1

Machine Stitches in Use
常用机缝线迹

Machine stitches are the basic stitches in tailoring and dressmaking. In a sense, no machine stitches, no modern garments. Every worker must master the skill. Machine stitches play an important role in sewing work. Machine stitches are permanent stitches made by sewing machines. It is sewing method that makes garment pieces and pieces close together. They are the most important stitches in tailoring and dressmaking; they are also the most useful stitches. We have no sewing without machines stitches. They are used to hold fabrics and parts of a garment together. There are many basic machine stitches in our tailoring and dressmaking. They are important and useful in our work. Let's learn some of them.

1. Top Stitches

Top stitches are most widely used stitching method. Top stitches are one of the most useful stitches we use in sewing. Top Stitches are most common and widely used in our tailoring and dressmaking. We must pay special attention to them. The fit and outline of a garment depend largely on the way in which the stitches are made.

The color of the thread to use depends on that of the material. Top stitches are a line of machine stitching on the visible side or right side of a garment to a seam. Usually the stitches are very close to the finished edges and are parallel to the finished edges.

2. Overlaid Stitches

Overlaid Stitches are other common and useful stitches. It is necessary for us to do them well. They are often used in tailoring and dressmaking. Overlaid stitches are a line of machine stitching alongside the turned edge and parallel to it. When stitching, turn over the seam allowance smoothly and stitch on the right side. The stitches must be made carefully, otherwise they will be bad look and leave the garment poor quality.

3. Reinforce Stitches

Reinforce stitches are used to reinforce some areas and some sections at right parts. Reinforce stitches are a line of machine stitching that strengthen an area of a garment such as pockets, side seams, back seams of trousers, underarm seams and belts. They are usually double line stitches at all right parts.

New Words and Expressions

basic	基本的
modern	现代的
master	掌握
permanent stitches	永久针迹
in a sense	从某种意义上说
fabric	织物
top stitches	明缉线，清止口
hold...together	将……并拢
parallel to	与……平行
edge	边缘
outline	外形
overlaid stitches	折边缝线
visible	可见的
fit	合体的
strengthen	加强
it is...that	正是……才
be widely used	被广泛应用
play an important role	起重要作用
right side	正面
common	常用的
alongside	沿着
turn over	折转
seam allowance	缝头
smoothly	平整地
otherwise	否则
reinforce	加固
underarm seam	腋下缝
back seam	后缝
area	区域
outline	轮廓
largely	很大程度上

Notes to the Text 1

1. Machine stitches are permanent stitches made by sewing machine. 机缝针迹是用缝纫机缝制的永久线迹。

2. They are used to hold fabrics and parts of a garment together. 它们用于固定服装零部件和织物。

3. The fit and outline of a garment depend largely on the way in which the stitches are made. 服装的外形和合体很大程度上取决于线迹的缝制方法。

4. Top stitches are a line of machine stitching on the visible side or right side of a garment to a seam. 明缉线是一行服装正面缝迹可见的机缝线。

5. Overlaid stitches are a line of machine stitching alongside the turned edge and parallel to it. 折边缝是一行沿着折转的缝头边缘并与之平行的机缝线。

6. Reinforce stitches are a line of machine stitching that strengthen an area of a garment such as pockets, side seams, back seams of trousers, underarm seams and belts. 加固线是一行在服装的某部位例如口袋、裤子边缝、后缝腋下缝和腰带处的机缝线。

Text 2

>> Seams and Seam Finishes
接缝缝合处理

In order to sew and stitch a garment perfectly and correctly, we must finish the raw edges of seams before making up garments. This is a very important and necessary step. We must not neglect this step. Ignoring this step or failing to do it well will leave a lot of quality problems in the clothes. The quality of a garment largely depends on it. If we can't do it well, the quality of a garment will be a failure. There are many ways to do the job. Each way is vital. We must make serious efforts to finish the seams. We have many ways to finish the raw edge of seams. Each way has its own function. The way we select to finish the raw edge of seams depends on the kind of fabric or material we are going to use, the type of garment, the need for a decorative and so on. The main purpose of a seam finish is to prevent the fabric from raveling, stretching, or curling.

Doing seam finishes needs patience and perseverance. If we work hard to do the seam finishes, we are sure we will learn a great deal from it. We usually have many ways to do seams and seam finishes, they are tacking, tailor's knot, back stitching, taped seam, edge stitch, French stitch, bound seam, flat fell seam and welt seam, crack stitch, lapped stitch and so on. Our seam allowances are usually 1cm for straight edges and 0.8cm for curves. Before making up a garment we should prepare the seam for the correct finish. We stitch the seam, trim the seam where it is necessary and press the seam.

New Words and Expressions

seam finish	缝迹加工处理
perfectly	完美地
correctly	正确地
neglect	忽视
ignore	不理会
fail to do	没能做
leave	留下
finish	处理
step	步骤
largely	很大程度上
failure	失败
vital	重要的
make serious efforts	认真努力
function	作用
select	选择
main	主要的
purpose	目的
patience	耐心
perseverance	毅力
a great deal	很多
tacking	加固缝，来回针
tailor's knot	套结
back stitching	倒钩针
taped seam	牵带缝
edge stitch	清止口
French stitch	来去缝
bound seam	包边缝
flat fell seam	明包缝
welt seam	暗包缝
crack seam	漏落针
lapped seam	搭缝
parallel to	平行
raw edge	毛边
decorative	装饰的
prevent	防止
ravel	散开
stretch	伸展
curl	卷曲
curve	弯曲
trim	修剪

Notes to the Text 2

1. In order to sew and stitch a garment perfectly and correctly, we must finish the raw edges of seams before making up garments. 为了完美准确地缝制服装，我们必须在制作服装前对毛边进行处理。

2. Ignoring this step or failing to do it well will leave a lot of quality problems in the clothes. 忽视这一步或未能做好这一步会在服装上留下很多质量问题。

3. There are many ways to do the job. Each way is vital. We must make serious efforts to finish the seams. 处理缝迹的方法有很多。每一种方法都很重要。我们必须认真努力地处理好缝迹。

4. Doing seam finishes needs patience and perseverance. If we work hard to do the seam finishes, we are sure we will learn a great deal from it. 处理缝迹需要耐心和毅力。如果我们努力做好这工作，我们相信我们会从中学到很多东西。

5. We stitch the seam, trim the seam where it is necessary and press the seam. 在需要的部位修剪、缝缉和熨烫缝迹。

>> Substitution Drills　替换练习 <<

1. Turn over the seam allowance smoothly and [stitch / press / tack] on the right side.

2. There are many basic [machine stitches / hand stitches / temporary stitches] in sewing work.

3. I asked if they had a cheap [skirt / shirt / jacket].

4. We use reinforce stitches to strengthen [the bask seams / the pockets / the underarm seam].

5. Stitch smoothly alongside the [turned edges / seam allowance / belt].

Exercises to the Texts 课文练习

1. 抄写并熟记本课短语和词组
2. 翻译下列短语

1）machine stitches
2）permanent stitches
3）be used to
4）close to
5）in a sense
6）right side
7）turn over
8）overlaid stitches
9）underarm seam
10）reinforce stitches
11）depend on
12）such as
13）bad look
14）pay special attention to
15）a line of
16）double line stitches
17）seam allowance
18）parallel to
19）raw edge
20）press the seam

3. 将下列短语译成英语

1）缝头
2）折转
3）正面
4）掌握
5）完美正确地
6）很多做某事的方法
7）缝迹
8）修剪缝迹
9）防止散开
10）漏落针
11）与……平行
12）服装的外形与合体
13）明缉线
14）边缘
15）明包缝
16）暗包缝
17）毛边
18）搭缝
19）倒钩针
20）套结

4. 翻译

1）沿着折转的止口缝缉。

2）服装的外形与合体很大程度上取决于线迹的缝制方法。

3）机缝针迹是永久性线迹。

4）加固线迹很重要。

5）折边缝很常用。

6）线的颜色取决于织物的颜色。

7）我们应特别注意明缉线。

8）学好机缝很重要。

9）我们必须认真努力地处理缝迹。

10）处理缝迹需要耐心和毅力。

11）这条缝迹必须和那条缝迹平行。

12）明缉线是一行服装正面可见的机缝线迹。

13）我们必须在服装的口袋，裤子的边缝、后缝用加固缝。

14）机缝线迹是男女装的基本线迹。

15）每个工人必须掌握各种机缝线迹。

16）机缝线迹在服装中起重要作用。

17）为了准确完美地缝好服装，我们必须做好缝迹处理。

18）我们从书本上学到了很多东西。

19）折转缝头在正面缝纫。

20）这工具是用来修理缝纫机的。

Conversation

>> Laying the Patterns on the Material 排料

A: I learned how to draw and make patterns of garments, and how to take measurements. I'd like to learn something about laying the patterns on the materials. Could you please tell me how to do it?

B: OK, now we have taken the measurements and made patterns, tried on, and adjusted to fit, next step we lay the patterns on the materials.

A: What are the points for attention before laying the patterns on the materials?

B: The general rules for laying the patterns on the materials are as follows: Fold the materials in half lengthwise, see that the selvage edges are exactly together. Study the material for nap. If there is a pile or one way patterns, all the pieces must be placed the right way up. If the material has stripes or checks, make certain that these stripes or checks will be matched at the seams.

A: Is there anything else should we pay attention to?

B: Yes. We should check the number of the pattern pieces and ensure all the marks and grain lines on the patterns are right and correct. We should pencil the grain lines on the patterns. Make sure the grain line lies on a straight thread of the material. Measure in from the selvage for accuracy.

A: What will happen if the grain line doesn't lie exactly on a straight thread of the material?

B: The garment will not hang correctly nor wear well, and the garments will out of shape.

A: There are many large pieces and small pieces in a garment. Which piece should we lay first?

B: We should lay all the large pieces first, and then fit the smaller pieces into the space left. It's economical.

New Words and Expressions

lay the material	排料	right way up	正面朝上
adjust	调整	stripe	条子
points of attention	注意点	checks	格子
general rule	一般规则	make certain	确信
as follows	如下	match at the seams	在缝处对齐
lengthwise	纵向地	pencil	用铅笔画
see that	务必	accuracy	正确
selvage edge	布边	straight thread	直线
study	观察	hang	垂势
nap	绒毛	fit into	插入，放入
pile	绒面	out of shape	走样，变样
one way pattern	一顺花型	economical	节约的
from the selvage	从布边	measure in	量进
stripes be matched	对条子	checks be matched	对格子

Reading Practice 1

All Kinds of Stitches 各种线迹

There are many kinds of stitches in tailoring and dressmaking, which including temporary stitches, permanent stitches, hand stitches and machine stitches. All these stitches are common and useful in our production. They are basic skills in sewing and stitching garments. We must master them and skillfully use them in operation.

Temporary stitches are generally larger and farther apart than permanent stitches. Temporary stitches are used to hold the pieces and parts of a garment together temporarily or transfer the fitting lines from patterns to material. They should be removed when the garment is finished. So it is important in needlework. Any hole and pinhole mustn't be left when the garments are finished. Sometimes pin will be used to hold the pieces instead of thread.

Permanent stitches are used where the stitches are necessary. Permanent stitches are those which hold parts of a garment together. This kind of stitches is either made by hand or by machine.

Hand stitches are made by hand. They are traditional skills. They are used where machine stitches are not desirable. Hand stitches used depend on the garment style, the fabric and place where the stitching is necessary.

Machine stitches are usually made by sewing machines, sometimes they are made by other machines. They are used to stitch permanent seams or finished edges by machine, such as collar edges and front opening edges. A line of machine is usually reinforced with backstitches at both ends.

New Words and Expressions

temporary	临时的	remove	拆
permanent	永久的	desirable	称心的
farther	更进一步	backstitch	来回缝
apart	相距	end	顶端
temporarily	暂时地	skillfully	熟练地
transfer	转移	hole	孔
pinhole	针孔	reinforce	加固
hold...together	将……缝合在一起	pin	别针
fitting line	净缝线		

Reading Practice 2

>> Temporary Stitches and Permanent Stitches 临时线迹和永久线迹

There are many kinds of sewing and stitching ways in making up garments. They can be divided into two kinds, temporary stitches and permanent stitches. They are basic and common in production. Each way of stitching has its own function. Each way of stitching is very important in production. They are the skills that every tailor and dressmaker must learn and master. Our job requires a lot of skill. One of them is this kind of stitching. As the name suggests, temporary stitches last only a short time in the production. When the garment is finished they will be removed. Permanent stitches exist for a long period. We have machine stitches and hand stitches in permanent stitches. We usually have hand stitches in temporary stitches. These two kinds of stitches play an important role in sewing work. Temporary stitches are generally larger and farther apart than permanent stitches. Hand stitches are made by hand. They are used to decorate a garment. Hand stitching is important in needlework. It is a traditional skill in needlework. We use our skill in production and show great skill in our clothes.

>> New Words and Expressions <<

divide into 分为
period 一段时间
decorate 装饰
use one's skill in 运用自己的技能
show great skill in 显示很高的技巧
function 作用
require 要求

Lesson Seventeen

Text 1

>> Hand Stitches 手缝线迹

Hand stitches are the basic skills in tailoring and dressmaking. They are common and important in our production. Hand stitches are traditional skills. Hand stitching contains many Chinese elements and culture. They are widely and commonly used in Chines gowns and robes. They reflect the Chinese culture. Hand stitches are a line of stitches made by hand. Mastering hand stitches is one of the basic skills for a tailor or a dressmaker. Hand stitches are also traditional skills in sewing. Hand stitching is passed down by generations of garment masters and designers, which contains the wisdom of masters and workers. Based on some Chinese elements, hand stitches achieve great popularity in the world and won large customers. In order to master hand stitches well, we need to practice for a long time. The stitches are passed down from generation to generation, so they are practical and valuable. The stitches are used to sew two pieces or parts of a garment together. Hand stitches include temporary stitches and permanent stitches. Temporary stitches are used to transfer the fitting lines from patterns to material, or to hold the parts together temporarily before stitching. They are a guide line for machining and for holding parts, such as pleats, hems and cuffs. When the garment is finished, hand stitches will be removed. We should use a fine needle with cotton thread so we will not leave any traces of the thread in the garment when the stitches are removed. Usually we use white cotton thread to make the stitches.

Hand stitches can also be used for permanent stitches. Hand stitches are used to stitch plain seams, top stitched seams, double stitched seams and so on. Hand stitches are often used for babies' clothes because they are soft. Hand stitches are used in traditional Chinese wear. Sometimes we use hand stitches for show. Hand stitches are seen on the right side of the garment. The outline of a garment depends on the way the seams are made. It is very important to make the hand stitches when you stitch a garment. We cannot be too careful when we are making the hand stitches.

New Words and Expressions

basic skill	基本技能
traditional skill	传统技能
contain	包含
be widely used	广泛使用
reflect	反映
based on	基于
Chinese elements	中国元素
for show	为了引人注目
valuable	有价值的
fine	细的(针)
guide	引导
leave	留下
cannot be too careful	怎么仔细也不为过
achieve	取得
wisdom	智慧
pass down	流传下去
generation	代
practical	实用的
trace	痕迹
plain	平缝
guide line	指导线
pleats	褶裥，褶皱
hems	底边
cuffs	袖头
remove	拆除
gown	袍
robe	披肩
plain seam	平缝
top stitched seam	明缉线
double stitched seam	双缉线
right side	正面
outline	轮廓线

Notes to the Text 1

1. Hand stitches are also traditional skills in sewing. 手缝线迹也是缝纫工艺中的传统技能。

2. Based on some Chinese elements. 基于一些中国元素。

3. The stitches are passed down from generation to generation, so they are practical and valuable. 这些针法一代一代流传下来，因此它们非常实用和有使用价值。

4. They are a guide line for machining and for holding parts, such as pleats, hems and cuffs. 它们是作为引导固定褶裥底边和袖头等零部件的机缝引导线。

5. ...will not leave any traces of the thread in the garment when the stitches are removed. ……当服装上的线拆除后不会留下痕迹。

6. The outline of a garment depends on the way the seams are made. 服装的外观取决于接缝的缝法。

7. We cannot be too careful when we are making the hand stitches. 当进行手缝时无论怎样仔细也不为过。

Text 2

Sewing Threads 缝纫线

In order to achieve the required result to the garment and to aid the workers during the making up garments. It is very important to select right and correct sewing threads before making up garments. Sewing threads are made from a wide variety of materials, such as wool, cotton, synthetic fibers. There are also many kinds of blended sewing threads. We have many threads in our tailoring and dressmaking. The threads we use in production are cotton threads, silk threads, synthetic threads and blended threads. We have threads in an assortment of color and counts.

Cotton threads are soft, warm and rough. Cotton threads are common in use. It is used widely in making up cotton garments. The disadvantage of cotton thread is that it shrinks greatly and is not very strong.

Silk threads are more soft. They have natural luster and elasticity. They are sometimes used for decorative stitching. They are used to sew silk garments.

Synthetic threads are strong. They are used to sew fabrics made from synthetic or man-made fibers.

Blended threads are made from synthetic or man-made fibers. We choose the right thread according to different fabrics.There are many kinds of synthetic threads in tailoring and dressmaking. They are common in use.

In addition, we also use a variety of hand sewing threads, such as double cotton thread.

New Words and Expressions

require result	要求的结果	fiber	纤维
aid	帮助	blended	混合的
a wide variety of	品种繁多的	natural	自然的
assortment	物品种类的组合	luster	光亮
count	（纱，线）支数	elasticity	弹性，弹力
synthetic	合成的	decorative	装饰的
disadvantage	缺点	man-made	人造的
shrink	缩水	common	常用的

Notes to the Text 2

1. In order to achieve the required result to the garment and to aid the workers during the making up garments. It is very important to select right and correct sewing threads before making up garments. 为了达到服装的要求和有助于工人们缝制服装。制作服装前挑选合适的缝纫线是非常重要的。

2. We have threads in an assortment of colors and counts. 我们有各种颜色和支数的缝线。

Substitution Drills 替换练习

1. Making [hand / machine / permanent] stitches is one of the basic skills for a tailor or a dressmaker.

2. | Hand stitches
Machine stitches
Permanent stitches
Temporary stitches | are also traditional skills in sewing.

3. In order to master | hand stitches
machine stitches
permanent stitches | well, we need to practice for a long time.

4. | Hand stitches
Machine stitches
Permanent stitches | are used to stitch plain seams, and so on.

5. | Hand stitches
Machine stitches
Permanent stitches | are seen on the right side of the garment.

Exercises to the Texts　课文练习

1. 抄写并熟记本课短语和词组
2. 翻译下列短语
   1）a pair of pants
   2）put on
   3）guide line
   4）sew two pieces together
   5）hand stitches
   6）basic skill
   7）Chinese gown
   8）a line of stitches
   9）achieve great popularity
   10）sew...together
   11）the outline of a garment
   12）right side
   13）master the skill
   14）fitting line
   15）a fine needle
   16）top stitched seam
   17）trace
   18）be widely used
   19）cotton thread
   20）pass down
3. 将下列短语译成英语
   1）反面
   2）留下痕迹

3）固定零部件
4）明缉线
5）拆除
6）双缉线
7）怎么仔细也不为过
8）服装的外观
9）基于
10）弹性
11）传统技能
12）基本技能
13）练习很长时间
14）棉线
15）合成纤维
16）装饰针迹
17）丝线
18）混纺线
19）自然光泽
20）除此之外

4. 翻译

1）这种针法在我们生产中很常用很重要。
2）手工针法很重要。
3）手缝线迹常用作引导线。
4）手缝线迹也可以用作永久针迹。
5）手缝是传统缝法。
6）我们应该每天练习手工针法。
7）熨烫时无论怎样小心都不为过。
8）请教我们怎样缝制这口袋。
9）我们有各种颜色和尺寸的衬衫。
10）棉线在生产中很常用。
11）服装正面可看到机缝针迹。
12）手工针迹是缝纫中的传统针迹。
13）制作服装前挑选合适的缝纫线是非常重要的。
14）有时候我们用手缝线迹是为了引人注目。
15）服装的外观取决于接缝的缝法。
16）缝制服装时无论怎样仔细也不为过。
17）这些中国元素是一代一代传下来的。
18）手工针法和机缝针法都是基本针法。
19）我们有各种颜色和支数的线。
20）这些技能非常实用而且有价值。

Conversation

>> The Different Features of Fabrics 各种织物的不同特点

A: We learn there are many kinds of fabrics and materials in garments. Do you know how many kinds of fabrics? What's the difference? Could you please tell me something about it? I'm very interested in it.

B: There are many kinds of fabrics in the market nowadays. The exact number is not clear. With the development of science and technology, many new varieties will appear. Some of fabrics are suitable for clothes, but some of them are not suitable for clothes. Some of them are out of date and some of them will be useful to clothes in the future. Now let's look at some samples of fabrics. Here you are.

A: Oh, how beautiful and colorful they are! I have never seen so many beautiful fabrics. Can they all be used to make clothes? How are they classified? How can we recognize them all at once?

B: Yes, they are divided into natural fabrics,blend fabric and synthetic fabrics. We have cotton cloth, woolen cloth, silk cloth in natural fabrics, nylon, polyester, rayon in synthetic fabrics.

A: What's the difference between natural fabrics and synthetic ones?

B: Natural fabrics are made of natural fibers. Cotton cloth is made of cotton. There is a great variety of cotton cloth such as plain cloth, poplin, checks, drill and so on. Chinese silk is well known all over the world, and it has a history of over three thousand years. Silk is popular in China and foreign countries. As we all know that silk feels soft and enriches the life of people. You get everything in silk, blouses, dresses and other clothes. Woolen fabrics are whipcord, gabardine, fancy suiting, palace, melton and so on. They are warm and stiff and smooth. Woolen fabrics are suitable for men's and women's suits. They are comfortable to wear. Synthetic fabrics are stronger than natural fabrics. They wash better and keep their shape well after washing. They absorb very little water, so they dry quickly. They are suitable for trousers, jackets, work clothes and so on.

New Words and Expressions

polyester	涤纶
rayon	人造纤维
plain cloth	平纹布
poplin	府绸
checks	格子布
drill	斜纹布
feel soft	手感柔软
enrich	丰富
blend	混纺
classify	分类
whipcord	斜纹呢
gabardine	华达呢
fancy suiting	花呢
palace	派力司
melton	麦尔登呢
absorb	吸收
keep shape	保形
variety	种类
recognize	识别

Reading Practice ❶

>> A Letter of Congratulations 一封祝贺信

Dear Lin Lin,

I received your letter the day before yesterday. I have been busy these days. So I fail to return your letter in time. Father, mother and I are quite well. We are glad to learn that you have made great progress in your study. We are proud of your success, but keep in mind that a person's life can't always successful. You should face bravely one challenge after another. Only those who face the challenges bravely may succeed. It will be hard for the student to design good fashions if you don't have innovative consciousness and innovative thinking. It will be difficult for you to design good fashions if you don't have a sound foundation.

How time flies! In about two weeks you will graduate from your garment school. Congratulations on your graduation! We are pleased to hear that you have made up your mind to prepare yourself to take part in the Garment Prize and Computer Contest in your city this summer. You study three and four hours every evening in addition to the work you do during the day. You are quite right. But you must take good care of yourself. All the possibilities should be considered before making a decision. Now more and more people have come to realize the importance of learning English and computer well. We wish you make full use of your time to do them well. With your talent, you are bound to make great progress so long as you work hard. We all know that not all intelligent students will become successful in the future. Other factors may influence their progress. If you need any help, just write to me. I'll do my best to help you. Please give my best regards to your classmates Li Tong and Mary.

Your sincerely

Wang pin

>> New Words and Expressions <<

keep in mind	牢记	those who	凡是……的人
bravely	勇敢地	graduate	毕业
challenge	挑战	graduation	毕业

innovative consciousness	创新意识	intelligent	聪明的
congratulation	祝贺	factor	因素
come to	渐渐开始	sound foundation	扎实的基础
make full use of	充分利用	innovative thinking	创新思维
give one's regards to	向……问候		

Reading Practice ❷

>> Preparation of Fabric 面料准备

In order to make clothes perfect, they will be neat and elegant without deformation in the future. Preparation of fabric is needed. It is important and necessary to understand fabrics before making up garments. Fabric is the carrier of clothes. Fabric is the foundation of clothing. Preparing fabrics is one of the most important procedures before making up garments. Sometimes fabrics will not hang correctly for a variety of reasons, in other words, the warp and weft of fabric is not vertical. If we cut out the garments and make up the garments, the garments will out of form and hang badly. We must block the fabric and insure the fabric of warp and weft correctly hang to a garment after it is made up. If the yarns do not form a 90° angle, the fabric should be blocked. Fold the fabric to form a true bias, then stretch the entire length of the fabric from selvage to selvage, pulling from the shorter end. Stretching on the true bias will line up the yarns squarely. To set the yarns, we should press the fabric on the wrong side with the iron heated to the correct temperature. The temperature of the iron must be well controlled. Otherwise, it will damage the expectation or fail to meet the design requirements. The work must be done patiently and carefully. In order to ensure good results, after block the fabric, the fabric should be hung up dry for several hours before layout. We mustn't use the improperly finished fabrics for garments. When these fabrics are used, the grain lines of the garments will be wrong. The garments will go out of form.

New Words and Expressions

perfect	完美的
elegant	优雅的
deformation	变形
carrier	载体
foundation	基础
procedure	步骤
warp	经线
weft	纬线
vertical	垂直
block	使归正
hang	垂势
yarn	纱线
to set the yarns	为了使纱线定型
damage	损坏
expectation	期待的效果
requirement	要求
bias	斜丝缕
stretch	张开，拉长
line up	使成行
squarely	端正地
go out of form	变形

Lesson Eighteen

Text 1

>> Introducing a Showroom in a Shoe Factory 介绍一家鞋厂的陈列室

This is a showroom in a shoe factory. Many kinds of traditional, modern, embroidery and ordinary shoes, slippers and sports shoes for men, women and children are on display. There are all kinds of sneakers. All the designs are very attractive. Many shoes are designed specially. Some are with Chinese elements the shoes achieved great success in markets and won large customers. Shoes made in China are cheap but in good quality, so they are very popular with the customers everywhere. It is here that businessmen, designers from all over the world and all parts of the country visit and have a look at the samples. The factory produces hundreds of thousands of shoes every year. Most of them are latest styles, some are traditional styles. They are made of cotton, woolen, synthetic fabrics and plastic. The vamp material is superb and durable. The shapes of the shoes produced in this factory are fit for different people. The workmanship and stitching are of top quality. The factory exports shoes to many countries and areas of the world. The shoes they produce win great popularity in foreign markets.

A shoe exhibition is being held in the showroom. A lot of businessmen and visitors have come, including the directors, designers, technicians and managers of foreign companies and firms. The Chinese businessmen and staff members are busy preparing the contracts and some documents. They should get everything ready before negotiation. Some foreign businessmen are looking at the exhibits and samples carefully, and some of them are discussing the styles, color, sizes and preference with the designers of the factory. Some businessmen and the staff members are having the contract drawn up. Some merchants are signing contracts with the salesmen of the factory. What a lively and busy exhibition!

New Words and Expressions

embroidery	刺绣的	workmanship	工艺
ordinary	普通的	win popularity	受欢迎
slipper	拖鞋	director	经理
sneakers	运动鞋	technician	技术员
sample	样品	firm	公司，商号
produce	生产	preference	爱好
be fit for	适合	draw up	草拟
traditional	传统的	merchant	商人
vamp	鞋面	sign contract	签合同
superb	华丽的	lively and busy	生动忙碌
durable	耐用的	document	文件
shape	使……成型	specially	特别地
negotiation	谈判	element	元素

Notes to the Text 1

1. The workmanship and stitching are of top quality. 工艺和线迹都是一流的。

2. Some businessmen and the staff members are having the contracts drawn up. 一些商人和工作人员正在起草合同。

Text 2

A Showroom 陈列室

This is a showroom in our garment factory. Many kinds of traditional Chinese gowns and fashionable modern clothes for men, women, and children are on display. Many garments are designed specially. After years of efforts and innovation our designers and workers have designed a variety of attractive and fashionable clothes. The designs of our products are more interesting than those of others. Some are with traditional Chinese elements. Look, the designs on the skirt are elegant. Our products enjoy great popularity in the world. Our factory enjoys a very high international repute. Our garments achieved great success in markets and won large customers. Garments made in our factory are cheap but in good quality, we have placed an emphasis on the quality of our products, so they are very popular with the customers everywhere. We continue to innovate, improve quality and increase variety. We designed and produced traditional garments and modern fashions to meet different people.

It is here that merchants and businessmen from all over the world and all parts of the country visit and have a look at the samples of all kinds of clothes. They talk about and discuss the styles, color, sizes and preference with the designers of our factory. Sometimes they draw up the contract and sign it after mutual negotiations, many transactions are concluded through mutual accommodation.

New Words and Expressions

gown 袍，中装
elements 元素
innovation 创新
design 花型
elegant 优雅的
repute 声望
place an emphasis on 强调
mutual 互相
transaction 交易
conclude 达成
accommodation 照顾，通融
innovate 创新
it is here that 正是在这里
increase 增加
variety 品种

Notes to the Text 2

1. After years of efforts and innovation our designers and workers have designed a variety of attractive and fashionable clothes. 我们的设计师和工人们经过多年的努力和创新设计出多种吸引人的流行服装。

2. Look, the designs on the skirt are elegant. 瞧，这裙子上的花型多优雅。

3. We continue to innovate, improve quality and increase variety. 我们不断创新，提高质量，增加品种。

Substitution Drills 替换练习

1. This is a showroom in a shoe / garment / clothing factory.

2. Many shoes / dresses / suits / jackets are designed specially.

3. The shapes of the shoes / trousers / shirts produced in this factory are fit for different people.

4. A shoe / garment / fabric exhibition is being held in the fair.

5. Some businessmen / directors / merchants and the staff members are having the contract drawn up.

6. The [workmanship / fabrics / silk dresses] and stitching are of top quality.

>> Exercises to the Texts 课文练习 <<

1. 抄写并熟记本课短语和词组

2. 翻译下列短语

1）draw up
2）superb
3）the shapes of the shoes
4）embroidery shoes
5）shoe factory
6）embroidery blouse
7）on display
8）attractive design
9）Chinese element
10）achieve great success
11）workmanship
12）staff member
13）sign a contract
14）win popularity
15）latest style
16）traditional style
17）be fit for
18）be busy preparing
19）lively and busy
20）place an emphasis on

3. 将下列短语译成英语

1）工艺
2）样品服
3）中国元素
4）签合同
5）中装
6）特别设计的
7）享有很高的声誉
8）观看样品
9）谈款式
10）起草合同
11）适合
12）大获成功
13）价廉质优
14）运动鞋
15）赢得客户
16）达成交易
17）相互通融
18）合同和文件
19）鞋面
20）使成型

4. 翻译

1）我们和外商签了合同。
2）他们正在草拟合同。
3）我们的工艺和质量是一流的。
4）这些运动鞋是为他们特别设计的。
5）他们厂生产的服装适合不同的人。
6）你和那公司签了合同吗？

7）这厂每年生产成千上万的夹克。
8）一个时装展正在那里举行。
9）所有的设计都非常吸引人。
10）所有的夹克都是特别设计的，有的含有中国元素。
11）正是在这里签订的合同。
12）工艺和缝纫都是一流的。
13）他们正在忙着准备合同和文件。
14）他们生产的不同鞋型适合不同的人。
15）我们生产的服装在外国市场上很受欢迎。
16）他们正在和设计师们讨论款式、尺寸。
17）一场鞋展正在举行。
18）我们达成了很多交易。
19）通过互相谅解我们签订了合同。
20）我们设计生产不同的服装满足不同人的需要。

Conversation

>> Some Common Seams 常用缝法

A: What does “seam” mean?

B: A seam is the line along with pieces of fabrics is joined by stitching. We can also say a seam is a line where two edges of cloth are turned back and sewn together.

A: Is it important in sewing?

B: Yes, it’s very important in sewing.

A: What’s the importance of seams?

B: The fit and outline of a garment depend largely on the way in which the seams are made. It is very important to be accurate when you are stitching the seams and to pay attention to the fitting line.

A: What type of seams should we use when we are stitching a garment?

B: The type of seam to use depends on the material and on the amount of wear and tear the garment will have to withstand.

A: How many types of seams we usually use in our sewing work?

B: In general there are four seams. They are plain sea, French seam, lapped seam and felt seam. Plain seam is a common seam in tailoring and dressmaking. With the wrong sides of the fabric outwards, stitch directly along the fitting line. Press the seam open. Two thirds of seams in a garment are plain seams.

A: How do we stitch French seam?

B: French seam is suitable for light fabrics when any other seams would be visible on the right side of the garment. With wrong sides of the fabrics together stitch along half seam allowance and trim to 0.3 cm. Then turn to the right side and stitch along the fitting line.

A: What's a lapped seam and how to stitch it?

B: A lapped seam is a seam that laps on the other, turn over the seam allowance smoothly to the wrong side. Place the turned piece right side outwards over the seam allowance of the other piece. Stitch along the fitting line.

A: And flat felt seam?

B: With two pieces of fabrics right together. One piece overlaps another stitch and press the seam to one side. Trim away the inner seam allowance stitch close to the folded edge.

New Words and Expressions

seam	缝法
join	缝合在一起
outline	外观
depend largely on	很大程度上取决于
accurate	精确的
fitting line	净缝
wear and tear	磨损
withstand	经受
plain seam	平缝
French seam	来去缝
lapped seam	搭缝
felt seam	包缝
outwards	朝外
press the seam open	分缝烫平
visible	看得见的
seam allowance	缝头
trim	修剪
right together	正面放一起
overlap	搭上

Reading Practice 1

Making a Dress for Myself
给自己做衣服

Mary was just straight from school. She wanted to look for a job to support herself. She was going to have an interview with a garment company. She had been thinking about how to make a good impression. She was well trained how to be a good designer in her school. Recently more and more people want to study abroad, but Mary was absorbed in learning how to design all day, hoping to realize her dream one day.

Now she can take the customer's measurements and according to the measurements to make patterns. And lay the patterns on the materials to cut out the garments. She was also well trained to be a good dressmaker. But of course appearance was also important. She had to look like a professional. She needed something that would look beautiful in the office. So she decided to make a beautiful dress for herself. She designed and drew a dress picture, and made patterns. After that she bought a piece of silk cloth and cut out the dress. Then she tried it on and made some alterations in the waist and chest. It was a princess style and with a peter pan collar on it. After standing in front of the mirror looking at herself, she was pleased with herself, because she did look beautiful. She was sure that dress would bring her a good luck. The company would certainly employ her. She was confident that she would get the job.

>> New Words and Expressions <<

interview	面试	confident	自信
impression	印象	appearance	外表
princess	公主	professional	专业

Reading Practice ❷

>> Designing a Dress 设计女装

Lucy was graduated from a garment school last year. She was graduated with first-class honors. Soon after graduating, she wanted to try her luck, and she applied for a job as a designer in a clothing company and had been accepted. She was admitted by a big clothing company. She wanted to leave a good impression on the interviewer. She wanted to design a dress for herself for her interview wear. She liked to be the center of attention. She was eager to get the job because she loves it so much.

She first drew fashion figure and pictures. She made some alterations again and again. Then she asked her teacher and classmates for advice. It was designed on the model of the latest style. She asked a designer to take her measurements and according to the measurements to make paper patterns and cut out the papers carefully. She chose the silk material for her dress. She laid the patterns on the material to cut out the dress. She pinned the pieces together and tried the dress on. After standing in front of the mirror looking at herself, she was pleased with the dress. She was confident that the dress would be perfect. She was sure the dress would set her figure off to advantage.

New Words and Expressions

graduate	毕业
with first-class honors	以优异成绩
apply for	申请
try one's luck	碰运气
be eager to do	迫切想做
center of attention	注意中心
fashion figure	时尚人物像
admit	允许
interviewer	主持面试者
interview	面试
ask sb. for advice	向某人征求意见
pin	用别针别
set her figure off to advantage	把她的身段衬托得更加优美

Lesson Nineteen

Text ❶

>> Fabric Preparation 面料整理

Fabric preparation is the first step in garment production. It is necessary and important to have an understanding of fabrics in order to assemble garments correctly. To get the best result of garments and prevent the garments out of shape after washing, most fabrics require preparations. That is, we must check the grain of the fabrics. Whether the grain of the fabrics is straight or not is important for a garment. If we use the wrong grain in a garment, the garment will not hang well. In addition, we must allow for the shrinkage of them so that the fabrics are suitable to the garment. Otherwise, the garment will shrink out of shape after washing. Usually we shrink the material before cutting it. Do it as follows.

1. Spread the fabrics wrong side up on a flat worktable.

2. Place an L-shaped square on the fabric. Align one side of the square with a selvage.

3. Draw a chalk line along the other side of the square so that the line is at a right angle to the selvage.

4. Cut along the chalk line.

5. Repeat at the opposite end of the fabric.

6. If the fabric is washable, pre-shrink it by immersing it in cold water for about two hours. Then gently squeeze the water, but do not wring.

7. If the fabric is not washable, use a steam iron on the wrong side of the fabric start ironing at the selvage until you have ironed the entire length of the fabric.

8. Put the material into an airy place to dry.

9. Lay the fabric flat on the work table carefully.

New Words and Expressions

prepare	整理，准备	along	沿
first step	第一步	angle	角
assemble	装配	repeat	重复
correctly	准确地	opposite	对面的
require	需要	check	检查
preparation	准备	washable	可洗的
grain	丝缕，纹理	immerse	浸入
allow	允许	gently	轻轻地
shrinkage	缩水（名词）	squeeze	挤压
shrink	缩水（动词）	wring	拧绞
out of shape	变形	entire	整个
spread	铺	pre-shrink	预缩
wrong side	反面	length	长度
place	放	airy	空气的
square	平方，直角尺	lay	放
selvage	布边		

Notes to the Text 1

1. To get the best result of garments and prevent the garments out of shape after washing, most fabrics require preparations. 为了使服装得到最好的穿着效果，防止洗后变形，大部分织物都需要整理。

2. ...we must check the grain of the fabrics. ……我们必须检查核对织物的直丝缕。

3. ...wrong grain ……丝缕不直

4. In addition, we must allow for the shrinkage of them so that the fabrics are suitable to the garment. 除此之外，为了使织物适合我们必须预留缩水长度。

5. Spread the fabrics wrong side up on a flat worktable. 将织物反面朝上平铺在平整的工作台上。

6. Place an L-shaped square on the fabric. Align one side of the square with a selvage. 将L型的直尺放织物上，使直尺的一边与织物的一边对齐。

7. If the fabric is not washable, use a steam iron on the wrong side of the fabric start ironing at the selvage until you have ironed the entire length of the fabric. 如果织物是不可洗的，在织物的反面从布边开始熨烫直至全部。

8. Put the material into an airy place to dry. 将面料放在通风处晾干。

Text 2

>> Layout and Cutout
铺料和裁剪

It is a very important step to lay the pattern pieces on the fabric. There are many rules and regulation in layout and cutting out clothes. We must keep them in our minds. If we neglect them, we'll be in trouble. It is known to all that carelessness can cause many mistakes. The layout of pattern pieces on the fabric is very important before cutting out garments. The position must be accurate and the straight grain line must be aligned with the wrap of the fabric when we lay the pattern pieces on the fabric. There is no doubt that checking the pattern pieces before layout has a great effect on the quality of the garments. We must check the number of pieces, parts of the garments. We should check the grain lines, marks on the pattern pieces. Every pattern piece should be marked grain line. The seam allowances should be added according to the method of stitching, type of fabric and type of garment. It is very important that the seam allowances are accurate. The size of the garment depends on them. The layout should be planned for the least amount of waste.

On the folded fabric, we should always plan the layout on the wrong side with the right sides together. When we do this job we must pay much attention to it. If we do something wrong there will be major mistakes. So we cannot be too careful when doing it. The fabric folded along the lengthwise grain or crosswise grain. The number and width of the pattern pieces determines the direction in which the fabric is folded. When we cutting out clothes, we should pay special attention to safety. Safety is the first. Without safety, production is impossible. We should not only pay attention to our own personal safety, but also the safety of our products.

New Words and Expressions

regulation	规则
carelessness	粗心大意
position	位置
accurate	准确
align with	对齐
there is no doubt	毫无疑问
have a great effect on	对……起好的影响
mark	标志
check	核查
layout	布局
grain line	直丝缕线
seam allowance	预留的缝份
accurate	精确的
folded	折叠的
the least amount of waste	最少的浪费
lengthwise	长度
major	大的
cannot be too careful	怎么仔细也不过分
width	宽度
determine	决定
crosswise	成十字的
personal	自身的

Notes to the Text 2

1. The position must be accurate and the straight grain line must be aligned with the wrap of the fabric when we lay the pattern pieces on the fabric. 当我们在织物上摆放样板时必须注意位置要准确，样板上的直丝缕线必须和织物的经线一致。

2. We should check the grain lines, marks on the pattern pieces. 我们必须核对样板上的直丝缕线和对齐标志。

3. The seam allowances should be added according to the method of stitching, type of fabric and type of garment. 应该根据织物种类，缝纫方式，以及服装的种类预留缝份。

Substitution Drills 替换练习

1. Spread the fabrics [wrong side up / right side up / smoothly] on a flat worktable.

2. It is necessary and important to have an understanding of [fabrics / hand stitches / machine stitches].

3. Usually we [shrink / iron / prepare] the material before cutting it.

4. The [dress / blouse / skirt / T-shirt] looks very nice indeed.

5. Everybody here seems to watch the [fashion / garment] show.

Exercises to the Texts 课文练习

1. 抄写并熟记本课短语和词组
2. 翻译下列短语

1）prepare the fabrics
2）L-shaped square
3）assemble a jacket
4）right angle
5）out of shape
6）pre-shrink
7）wrong side
8）get the best result
9）the grain of the fabric
10）be suitable for
11）the length of the fabric
12）have an understanding of
13）flat worktable
14）wrong grain
15）be confident that
16）as follows

17）chalk line
18）airy place
19）opposite end
20）folded fabric

3. 将下列短语译成英语

1）布边
2）反面朝上
3）预留缩水
4）变形
5）预留的缝头
6）长度
7）第一步
8）毫无疑问
9）挤压
10）拧绞
11）轻轻挤压
12）检查直丝缕
13）织物的长度
14）织物的宽度
15）将织物浸入水
16）在布边熨烫
17）（衣服）垂势好
18）丝缕不直
19）排料
20）浪费最少

4. 翻译

1）了解织物很重要。
2）织物的丝缕是否直对服装很重要。
3）裁剪之前我们应将织物缩水。
4）在织物的反面排样。
5）对这种织物我们必须预留缩水尺寸。
6）沿着划线裁剪。
7）用蒸汽熨斗在织物的反面熨烫。
8）裁剪前应仔细核查衣片数量。
9）我们必须检查核对织物的直丝缕。
10）必须检查衣片的数量和对刀标记。
11）裁剪之前核查样板的数量是必要的。
12）防止衣服变形是很重要的。
13）将织物反面朝上铺在桌面上。
14）将织物浸在冷水中两小时。
15）位置必须准确。
16）在织物的反面划一条线。
17）了解手工针法是非常重要的。
18）通常布料裁剪之前先缩水。
19）这件大衣看起来的确很漂亮。
20）摆放样板前必须核对样板的直丝缕线和对齐标志。

Conversation

>> Making Up a Garment 缝制服装

A: What should we do before making up a garment?

B: We should understand some principles before making up a garment or parts of a garment, and we should follow these principles in order to produce the garment in the minimum amount of time and with an obvious degree of profession.

A: What's the first principle should follow?

B: After cutting, the sections are arranged in bundle for easy assembly, usually in the order of making. Trimmings such as zips and stay tapes are included to prevent any interrupting during assembly.

A: What's the second should be considered?

B: Reinforcement is necessary at points of the garment which are expected to take strain. For example, pocket mouths can be used in the form of stay stitching to stabilize edges that are likely to stretch. The hang and the fit of a garment depend largely on it.

A: What's the next?

B: Select seam type. This is important factor when making up a garment. The seam may be a part of original design. It serves the purpose of joining sections to achieve the expected result.

A: Is there anything else we should learn?

B: We should know that each stage becomes preparation for the next. We must keep in mind that "accuracy and efficiency" at all time.

>> New Words and Expressions <<

principle	原则	in the order of	以……的顺序
trimmings	辅料	degree of profession	专业水平
minimum	最少	arrange	包扎成
obvious	明显的	sections	各种衣片
easy assembly	方便流水作业	include	包括
interrupt	中断，打断	prevent	防止
stay tape	牵带	take strain	承受压力
consider	考虑	be expected to	可能会
reinforcement	加固	at points of	在某些……点
take strain	拉紧	pocket mouth	袋口处
in the form of	以……的形式	stay stitching	固止线迹
purpose	目的	stabilize	（使）稳定，（使）稳固
efficiency	效率		
be likely	可能	factor	因素
follow	遵循		

Reading Practice ❶

>> Garment Fabrics 服装面料

There are many kinds of Chinese traditional fabrics. They are cotton cloth, woolen cloth, silk, synthetic cloth, blended cloth, homespun cloth and others.

Cotton cloth is made of cotton. There is a great variety of cotton cloth such as plain cloth, poplin, canvas, checks, drill, and corduroy and so on. Cotton garments enjoy an increasing demand.

Chinese silk is well-known all over the world, and it has a history of over two thousand year. Silk not only enriches the life of the people, but also earns much money for our country every year. Silk is popular in China and foreign countries and it is popular with the customers of the world. As we know that silk feels soft. Silk clothes are elegant, beautiful and attractive. Today you can get almost everything in silk, such as shirts, blouses, dresses and other clothes. The range of colors, patterns and textures is wide enough to suit everyone's tastes and needs. Silk fabrics are suitable for garments, especially brocade, georgette and crepe.

Woolen fabrics are whipcord, gabardine, fancy suiting, palace, melton and so on.

Homespun cloth is also one of our special products. It is wholly hand-woven cloth. It is tough, but soft and porous and it is made traditional styles and colors. So it is comfortable to wear.

>> New Words and Expressions <<

homespun	家纺的
poplin	府绸
corduroy	灯芯绒
enrich	使丰富
texture	结构
brocade	锦缎
georgette	乔其纱
crepe	绉绸
elegant	优雅
whipcord	马裤呢
gabardine	华达呢
fancy suiting	花呢
palace	派力司
melton	麦尔登呢
hand-woven	手工织造的
tough	结实的
porous	多孔的
an increasing demand	需求在增长
attractive	吸引人的

Reading Practice ❷

>> The Marks on Paper Pattern 样板上的标记

In order to ensure that the size of clothing is accurate and all parts are sewn in place, various alignment marks must be marked on the pattern pieces of clothing. This is a job that needs to be taken seriously. Without the alignment mark, the quality of clothing will be greatly affected. Their role is particularly important in the mass production of clothing. There are many kinds of marks on paper pattern. They are the positions of buttonholes, darts, balance marks, tucks, fitting lines, places to fold, fold lines, straight grain lines and so on. They are the positions we should pay much attention to in the operation. Their importance cannot be too much emphasized. We can be too careful in the operation. Here "in the operation" includes we should be careful in making paper patterns, cutting out paper patterns and stitching the garment. In a word, we should know the importance of the marks on paper patterns.

We perforate darts and tucks with small holes on paper patterns. We mark the positions of pockets and buttonholes with small holes. We mark the grain line with printed lines on paper patterns.

Balanced marks are small notches cut out of the edges of the pattern. The notches on one piece of pattern will match on another piece and show that the two pieces are to be joined together.

>> New Words and Expressions <<

ensure	确保
accurate	准确
sewn in place	缝制到位
alignment mark	对准标志
seriously	认真地
be greatly affected	受到很大影响
role	作用
particularly	尤为
mass	批量
fold line	折叠线
balance	平衡
fitting line	净缝线
operation	操作
emphasize	强调，重视
perforate	打孔
notch	切口
be joined together	拼在一起

Lesson Twenty

Text 1

>> Common and Useful Stitches 常见实用线迹

There are many useful stitches in tailoring and dressmaking. Some of them are traditional and some of them are changed with the time. Many basic stitches are with Chinese elements and reflect the Chinese styles. Basic stitches are not only used for holding and sewing garment parts and pieces, they can also be used for adorning dresses and children's wear. Some of them are innovative, outstanding style, eye-catching and pleasing to the eye. Now let's introduce some common stitches. They are herring-bone stitches, blind hem stitches, baste and running stitches.

1. The Herring-bone Stitch

The herring-bone stitch is worked from left to right, pick up one thread close to the hem edge, and then cross over diagonally into the garment, then pull the thread through. Take a small stitch in the hem only 0.1-0.2 cm down from the edge and 0.1-0.2 cm of the right of the previous stitch. Each with a fastening stitch. This stitch is often used for lined garment where the raw edge is left unfinished.

2. Blind Hem Stitch

Insert the needle into the folded edge of the hem, pick up one thread in the garment directly below the edge of the hem, slant the needle a little away and pick up the hem again. This stitch is used where the hem is held closely and stitches should show as little as possible on both the right and wrong sides of the garment.

3. Baste

Baste stitches are made by hand, and are removed when the garment is finished. Choose a needle which can be threaded easily and thread it with long double tacking cotton as the tacking cotton is fluffy and stays in the material.

4. Running Stitch

Running stitch is the basic stitch. Insert the needle from the wrong side of the garment (fabric) and weave the needle in and out of the fabric several times with knotted thread. Then pull the thread through.

New Words and Expressions

be changed with the times	随着时间而变化	innovative	创意新颖
element	元素	outstanding style	风格突出
eye-catching	夺人眼球	pleasing to the eye	令人赏心悦目
fastening	打结的	unfinished	未处理的
common	常用的	reflect	反映
adorn	装饰	end	末尾
basic	基本	fasten	打结
herring	鲱鱼	raw	粗糙的
herring-bone stitch	三角针迹	insert	插入
blind	暗的	fold	折叠
hem	底边	slant	使倾斜
blind hem stitch	底边暗缝线迹	remove	拆除
baste	扎缝	tack	扎
running stitch	通针	fluffy	多绒毛的
diagonally	对角地	weave	波动
previous	以前的	knot	打结
lined	有里子的	be threaded easily	容易穿线
raw edge	毛边	directly below	紧贴下面
double tacking cotton	双股扎线	pick up	挑起
fluffy	有绒毛的	pull through	抽出
stay	滞留		

Notes to the Text 1

1. ...pick up one thread close to the hem edge, ... ……紧靠底边挑起一根线，……

2. Each with a fastening stitch. 每针打个结。

3. The herring-bone stitch is worked from left to right, ... 三角针是从左往右进行的，……

4. Running stitch is the basic stitch. Insert the needle from the wrong side of the garment (fabric) and weave the needle in and out of the fabric several times with knotted thread. Then pull the thread through. 通针是基本针迹。将针穿线打结从服装的反面插入在织物上正反面等距缝制几针，然后将线抽出。

Text 2

>> Temporary Stitches 临时线迹

Temporary stitches are traditional stitches in sewing and stitching garments. It is a very useful method in production. It is a specialized and traditional craft skill. This kind of stitch looks ordinary, but it plays a very important role in production. Temporary stitches are considered essential to the skill of any garment worker. It could play a role permanent stitches couldn't play in making up a garment. Every tailor and dressmaker should master the skill. There are many useful temporary stitches in tailoring and dressmaking. It is very important and necessary for people to use these stitches in operation and production. Their role, however, cannot be underestimated. They are used to hold or join some pieces of garment or parts temporarily. Temporary stitches will be removed when the garment is finished. Temporary stitches are generally larger and further than permanent ones. We usually use cotton thread or other thread in operation. We should not use knots so that the they are removed easily. We should not leave the traces of thread in the garment. We mustn't leave any holes in the fabric. If we leave any traces or any holes in the garment, it will affect the quality and beauty of clothes. Temporary thread should contrast slightly in color with the fabric.

There are many kinds of temporary stitches in operation. They are even basting, uneven basting, slip basting, machine basting and pin basting. People usually decide which way to choose according to the style, production process and the fabric of clothes.

New Words and Expressions

specialized skill	专门技能	contrast	对比
craft skill	手工技能	even	均匀的，对称的
play a role	起作用	uneven	不均匀的，不对称的
join	钉在一起	slip	滑动
further	更远	decide	决定
knot	结	production process	生产方式
trace	痕迹		

Notes to the Text 2

1. It is a specialized and traditional craft skill. 这是一种特殊的传统手工艺。
2. Their role, however, cannot be underestimated. 它们的作用无论如何不能被低估。
3. Temporary thread should contrast slightly in color with the fabric. 临时针迹线应该和面料的颜色稍微有点反差。
4. People usually decide which way to choose according to the style, production process and the fabric of clothes. 人们通常根据服装的款式、生产过程和面料来决定选用哪一种方式。

Substitution Drills 替换练习

1. Many [basic / baste / running] stitches are with Chinese elements and reflect the Chinese styles.

2. [Baste / Hand / Temporary] stitches are made by hand, and are removed when the garment is finished.

3. Running / Baste / Blind hem stitch is the basic stitch.

Exercises to the Texts 课文练习

1. 抄写并熟记本课短语和词组
2. 翻译下列短语
   1）changed with the time
   2）garment parts
   3）children's wear
   4）Chinese elements
   5）common stitches
   6）traditional craft skill
   7）running stitches
   8）pick up
   9）from left to right
   10）close to the hem edge
   11）raw edges
   12）knotted thread
   13）stay in the material
   14）pick up one thread
   15）take a small stitch
   16）previous stitch
   17）lined garment
   18）uneven
   19）pull through
   20）pleasing to the eye
3. 将下列短语译成英语
   1）三角针
   2）通针
   3）底边暗缝线迹
   4）对角的
   5）手工技能
   6）痕迹
   7）有反差
   8）抽出
   9）双股扎线
   10）假缝线迹
   11）容易穿线
   12）折叠的边
   13）反面
   14）织物的反面
   15）几次
   16）服装的反面
   17）专门技能
   18）起作用
   19）紧靠底边
   20）装里子的服装
4. 翻译
   1）三角针和通针都是基本针法。
   2）在织物的反面仔细熨烫。
   3）许多基本线迹都带有中国元素。
   4）从织物的反面插入针。
   5）在服装的底部反面挑起一针。
   6）服装的正反面应尽可能不露线迹。

7）为了学好针法，我们应该练习。
8）掌握针法很重要。
9）这种针法是从左往右进行的。
10）扎缝是手工进行的。
11）服装完成后这些线迹将被拆除。
12）通针和底边针法都是基本针法。
13）我们不应该在衣服上留下痕迹。
14）临时针迹一般比永久针迹长。
15）临时针迹线应该和面料的颜色稍微有点反差。
16）我们已经学习了三角针、暗针、底边通针。
17）学好这些针法是重要的必要的。
18）手工针法在生产中非常有用。
19）临时线迹起着永久针迹起不到的作用。
20）生产中我们通常用棉线缝临时线迹。

Conversation

How to Draw a Skirt
如何绘制裙子结构图

The Specifications of the Patterns

Positions	Length	Waist	Hip	Waist Belt
Specifications	68	74	94	3

A: Mr. Wang, we have learned how to take the measurements. What should we do next? Could you tell us something about it?

B: Now let me show you how to draw a skirt. We first draw front piece then back piece, in the end we draw other parts such as waist belts and so on. Suppose the length of the skirt is 68 cm, the waist is 74 cm, the hip is 94 cm.

A: Which line shall we draw first?

B: Now, the front piece of the skirt. Let's draw the first line. 1. It's the basic line, it's also called front center. This line is very important. This line is the basis of the whole drawing. If this line is not drawn correctly, the whole drawing will fail in the future. It must be parallel with the selvage of the fabric. 2. Next we draw top line and bottom line (also called the length of the skirt). These two lines must be right angle with basic line. The bottom is 65 cm (skirt length minus 3 cm that is 68 - 3=65 cm). 3. Hip line: 1/10 hip measure +1 cm=10.4 cm. We measure it from top line. 4. Hip (side seam) 1/4 hip measure=23.5 cm. This line must be parallel with front center.

A: Shall we draw back piece?

B: Yes. The top line, bottom line hip line are equal to the front piece. Back center: This line must be right angle with top line. The width of back piece: 1/4 hip measurement = 23.5 cm. Front dart width = 5.5 cm, front dart length =11 cm. Back darts width =5.5 cm, back dart length =13 cm. Side seam: 2 cm from width of hip. Then we draw a curved line at the side seam.

New Words and Expressions

specification	规格
position	部位
front piece	前片
back piece	后片
waist belts	腰带
suppose	假如
the length of the skirt	裙长
front center	前中线
fail	失败
basis	基础
correctly	正确地
be parallel with	与……平行
selvage	布边
be right angle with	与……成直角
minus	减
hip measure	臀围尺寸
side seam	侧缝
front dart width	前省道宽
back dart width	后省道宽
curved line	弧线
drawing	图纸

Reading Practice 1

A Machinist 一位机修工

Tom was a tailor. He was very interested in all kinds of sewing machines since he was young. He worked in a clothing factory. He was very interested in repair of various sewing machines. He disassembled them, installed them again, and then disassembled them for repeated observation. He believed that nothing could stop him from achieving his ambition. At the age of twenty, he wanted to become a machinist to repair all kinds of sewing machines. His father tried to persuade him to stay in his department store. But Tom wouldn't think of it. He had made up his mind to learn to repair all kinds of sewing machines. He went to a big city by bike and began seeking a job. He went to several places, but didn't find a right job. At last he found a job in a sewing machine shop. He was satisfied with the job, but he soon ran into a problem. His room and meals at a boarding-house would cost 1,000 yuan a month. This was a big sum of money for him. To earn extra, he began washing dishes in a restaurant in the evening. He had to work 16 hours a day on his two jobs, but he didn't mind. He enjoyed what he was doing. Tom learned everything he could about repairing sewing machines. Some time later, he went to work for a company that produced all kinds of sewing machines. There he worked hard and learned a lot from the workers and technicians. He firmly believed that with his talent and hard work, he was bound to become a machinist. Before long, he was made foreman of the company. Later he went to a college to learn machine engineering and how to design and produce sewing machines. After many years of hard work, finally he became an engineer of the company.

New Words and Expressions

observation	观察	install	安装
disassemble	解开，分解	repeated	反复的
machinist	机械师	stop...from doing	阻止……做
in the repair of	修理	earn	挣钱
persuade	劝说	sum	一笔
seek a job	找工作	foreman	领班
board-house	寄宿屋	achieve	达到
ambition	目标	believe	相信

Reading Practice 2

A Successful Designer 一位成功的设计师

Mary was a dressmaker. She graduated from an art school. She finished her education at a garment vocational school. She was vocationally educated in a famous garment company. She learned how to draw clothes pictures, how to take measurements, how to make paper patterns, how to lay the patterns on the fabrics, how to cut out a garment and so on. She learned all kinds of sewing and stitching skills. She had great skills in handwork. She also learned several traditional sewing skills. The skills are very useful and important in making traditional clothes. She was very talented in sewing and designing clothes. She showed remarkable skill in clothes.

Now Mary worked as an assistant designer in a famous clothing company. Her good command of English and design enabled her to have an advantage over the other staff members in the company. She wanted to become a famous designer. She made innovations, updated design concept, studied hard on clothing production technology. She learned from the designers in the company and conducted some market researches before designing every garment. She accessed to a variety of information on the Internet. She was sure that she was bound to design something perfect with her talent. The product that she designed was finally got the recognition of the market.

>> New Words and Expressions <<

be vocationally educated 受过职业培训
be very talented in 在……很有天赋
good command of 熟练掌握
have an advantage over 比……略胜一筹
have great skills in handwork 手工很巧
remarkable 显著的
show remarkable skill in 显示卓越才能
assistant 助理
innovation 创新
update 更新
concept 观念
conduct 进行
research 研究
access to 获得
recognition 认可

Lesson Twenty-One

Text 1

>> Interfacing 衬

Interfacing is a piece of fabric sewn beneath the facing, breast, shoulders and some areas of a garment, usually at the inside of the collar, chest, breast, and lapel. It gives shape and firmness. Interfacing is of great use for garments. It is as important as bone to a man. It serves as a foil to a garment. It is used to support the garments and to give the good look and attractive appearance. There are many kinds of interfacings in tailoring and dressmaking.

1. Fabric Interfacing

Fabric interfacing is a strong highly resilient fabric made of a mixture of cotton, wool and hair. It is called foundation canvas. It is the most useful interfacing in tailoring and dressmaking. The interfacing is used on a garment to help shape and support the fabric, and to keep the garment from stretching, creasing or sagging.

2. Haircloth Interfacing

It is a wiry, extra-resilient interfacing fabric made of a mixture of strong cotton fibers, linen and tough horsehair. It is used to support and reinforce the chest and bust and shoulder areas of a garment. Haircloth interfacing is commonly used in superior quality clothes. It is high quality and expensive.

3. Cotton Interfacing

Cotton interfacing helps some sections of a garment shape and firmness. We usually use mull, sheeting or tailor's canvas in tailoring and dressmaking. It is used for collar, cuff, pocket and some areas. Mull interfacing is the softest and most durable.

4. Non-woven Interfacing

Non-woven interfacing is made from thermoplastic fibers or chemicals. These are synthetic fibers which melt and fuse together with the garment section when heat is applied. It is used for synthetic fabric garments.

New Words and Expressions

interfacing	衬	bone	骨骼
beneath	下边	crease	褶皱，起皱
breast	胸	sag	陷下
lapel	驳头	haircloth	马鬃衬
give shape	成型	superior	优越
firmness	坚固	wiry	强韧的
serve as	用作	tough	坚韧的
foil	衬托物	horsehair	马鬃
support	支撑	mull	细软布
attractive appearance	吸引人的外表	sheeting	细布
resilient	有弹力的	tailor's canvas	厚衬布
mixture	混合	non-woven	无纺的
foundation	基础	thermoplastic	热缩的
keep from...doing	阻止	chemicals	化学品
canvas	帆布，粗布	melt	熔化
shape	成型	fuse	融合
stretch	延伸	apply	应用
be of great use	很有用		

Notes to the Text 1

1. It gives shape and firmness. 它起到定型挺括的效果。

2. It serves as a foil to a garment. 用作服装的衬托。

3. ...keep the garment from stretching, creasing or sagging. ……防止服装延伸、起皱和塌陷。

4. It is a wiry, extra-resilient interfacing fabric made of a mixture of strong cotton fibers, linen and tough horsehair. 这是一种强力棉、麻和坚韧马鬃混合的强韧的具有超强弹力的内衬。

5. These are synthetic fibers which melt and fuse together with the garment sections when heat. 加热后，这些合成纤维和服装的部位黏合在一起。

Text 2

>> Business Negotiation 商务谈判

A: Hello, nice to see you. Welcome to our company. Let me give you my business card. It is a pleasure to have a chance to talk with you.

B: Hello, nice to see you, too. This is my business card.

A: You have visited our factory and workshops. What do you think of our factory and workshops?

B: Your factory is big, clean and modern. I was deeply impressed by the labor enthusiasm of the workers. I enjoy the fine workmanship and good quality of your products. I wish to enter into business relationship with you.

A: Thank you. I'm glad to hear that. We are engaged in both export and import business. We produce all kinds of clothes. We export a large quantity of garments to many countries. Our exports are well received in the world market. We can also produce according to customer's order. These are the samples of our new series of products.

B: I find the samples quite satisfactory. I'd like to place an order with you for 200 men's shirts for trial order. I wish that you could quote us your lowest price.

A: Quality is king. Our products are of top quality, as you have seen. As the price is reasonable, we believe it is acceptable to you.

B: As the market is weak, your price is found to be on the high side. Since your price is on the high side, it is not acceptable for us to accept.

A: Our price is quite reasonable, which has been accepted by other customers. We have concluded considerable business with other customers at this price. We can offer you a discount of 5%.

B: Thank you. We are glad we have greatly improved both business and friendship by joint efforts.

A: Now let's have a close look at the details of the contract. Have you read the contract carefully?

B: Yes, I have. Let's sign the contract.

New Words and Expressions

negotiation	谈判
enter into business relationship	建立业务关系
engage in	从事
business card	名片
be deeply impressed	印象深刻
enthusiasm	热情
workmanship	工艺
place an order with you for	向你们订……的货
conclude	达成
sample	样品
new series of	新系列
trial order	试订
quality is king	质量是硬道理（质量第一，质量为王，质量最重要）
weak	不景气
on the high side	价格偏高
offer you a discount of	给你个……的折扣
by joint efforts	共同努力
the details of	……的细节
sign the contract	签合同

Notes to the Text 2

1. I enjoy the fine workmanship and good quality of your products. I wish to enter into business relationship with you. 我喜欢你们产品的精巧工艺和优异质量。我希望同你们建立业务关系。

2. We are engaged in both export and import business. 我们从事进出口贸易。

3. As the market is weak, your price is found to be on the high side. 由于市场不景气，你们的价格有点偏高。

4. We have concluded considerable business with other customers at this price. 按此价格，我们已与其他客户达成相当多的协议。

Substitution Drills 替换练习

1. Interfacing / Pressing / Stitching / Grain line is of great use for garments.

2. It is used to support and reinforce the breast / chest / waist of a garment.

3. It is used to support the garments to give / show / appear the good look and attractive appearance.

Exercises to the Texts 课文练习

1. 抄写并熟记本课短语和词组
2. 翻译下列短语

1）good look
2）shape and firmness
3）some areas
4）highly resilient
5）give shape
6）serve as
7）fabric interfacing
8）haircloth interfacing
9）keep...from
10）extra-resilient
11）tailor's canvas
12）superior quality
13）trial order
14）as important as
15）a mixture of
16）most durable
17）be used for
18）synthetic fiber
19）together with
20）by joint efforts

3. 将下列短语译成英语

1）非织造布
2）细布衬
3）吸引人的外表
4）防止延伸
5）驳头里面
6）服装的衬托
7）支撑服装
8）加强胸部
9）有机会做某事
10）名片
11）有助于成型
12）起定型效果
13）马尾衬
14）领子的里面
15）留下深刻印象
16）建力贸易关系
17）从事
18）精致的做工
19）新系列
20）向我们报最低价

4. 翻译

1）衬起支撑服装的作用。
2）无纺衬由热熔纤维制成。
3）衬对服装很重要。
4）衬用作服装的衬托。
5）棉布衬用在服装的领子、袖头和口袋处。
6）马尾衬质量好价格高。
7）衬在服装中很常用。
8）我们必须防止服装起皱。
9）他们希望同我们建立业务关系。
10）衣衬是用来支撑衣服的。
11）我们已和他们公司达成了交易。
12）质量是硬道理。
13）衬起到定型的作用。
14）我喜欢你们产品的精巧工艺和优异质量。
15）我们从事进出口贸易。
16）我很高兴有机会向你学习。
17）你认为我们的连衣裙怎样?
18）我们的连衣裙在国际市场上很受欢迎。
19）她仔细看过合同了吗?
20）我对这质量相当满意。

Conversation

>> A Garment Trade Fair 服装交易会

A: Welcome to our stand! What can I do for you?

B: Thank you! I'd like to have a look at the clothes you are showing.

A: This is my business card. What's your name?

B: Thank you! My name is Wu Ling. I'm from ABS Company. This is my colleague,

Mr. Fang. It is said that your exhibits are very popular. People say your products are very attractive and novel. Some of them seem to be of the latest style. We are interested in your clothes. I want to see your products and designs for myself.

A: Thank you for your interest in our products. Are you interested to have a look at the sample book and the well printed product catalog? We believe our products will satisfy you.

B: I wonder if you can give us some more information about the products you are showing.

A: OK, we have now adopted natural fabrics for materials and the trimmings we used are of top quality. Look at the workmanship of these clothes. It's excellent. We have set a high standard for the quality of our clothes. The workers in our factory have much experience in the production of garments. Our export clothes are getting more and more popular in the world market. Our products enjoy great popularity in Europe. We are sure you will be satisfied with our goods.

B: What you said acquaints us with your company and your products, but we have no dealings with you before, this is the first time we have deal with you. Would you please make arrangements for us to visit your company and look at your assembly line? We hope this will not cause you inconvenience. We hope you will be able to meet our request. Would you please give us an illustrated catalog and price list? We will take these documents back to talk it over with our managers. We will give you a definite reply next week. And we will make an arrangement for further discussion the transaction. We are willing to cooperate with you in this line.

A: Thank you! We are looking forward to cooperate with you.

>> New Words and Expressions <<

trade fair	交易会
stand	摊位
adopt	采用
standard	标准
get popular	受欢迎
enjoy popularity	享有声望
acquaint	了解
dealing	交往
see your designs	看看你们的设计
make arrangement for	作出安排
novel	设计新颖
deal with	打交道
inconvenience	不便
meet our request	满足我们的要求

illustrated	有插图的	line	行业
catalogue	目录单	cooperate with	与……合作
definite	确切的		

Reading Practice ❶

>> Measurements for Bodice 女装上衣量体

Women's clothing measurement is a more delicate work. The measurements should be more careful and accurate. A little carelessness, the clothes will be out of shape in the future.

Measurements for bodice are taken in the following way:

1. Bust

Take this measurements round the fullest part of the bust with two fingers inside the tape measure. Keep the tape high up under the arms at the back.

2. Front Width

It is taken across the front where the sleeves are set in and at a level of 10 cm below the nape of the neck.

3. Front Length

Take this from the neck end of the shoulder to the center of a tape tied round the waist.

4. Waist

It is taken round the natural waist. Push the measure well down.

5. Black Width

It is taken across the back from where the sleeves are set in at a level of 10 cm below the nape bone.

6. Back Length

It is taken from the nape bone to the center back of a tape tied round the waist.

7. Sleeve Length

Take this from the top of the shoulder, over the bent elbow and up to the wrist.

8. Sleeve Width

Take this round the developed biceps muscle and add 10 cm for ease.

9. Wrist

It is taken round the wrist for a sleeve with an opening. For a sleeve without an opening take this measurement round the clenched fist.

New Words and Expressions

bodice	（女装）大身部位	biceps	二头肌
delicate	精细的	muscle	肌肉
bust	胸围	wrist	手腕
nape	颈	opening	开衩
bone	骨头	clenched	紧握
bent	弯曲的	fist	拳头
elbow	肘部	a little carelessness	稍有疏忽
developed	发达的	out of shape	走样

Reading Practice 2

The Development of Sewing Machines 缝纫机的发展

Now we all know that a sewing machine is an expensive piece of equipment in our tailoring and dressmaking. But before the 1800s, people used hand needles to sew clothes. Like most important inventions, the sewing machine was the result of an important need of the time. The demand for more speed and increased production in the manufacture of garments in the early industry revolution brought about the change of stitching. The first serviceable sewing machine was developed by a poor French tailor, who invented one-thread sewing machine in 1830; the machine was six times faster than a skilled worker could sew by hand.

In the next few years some inventors produced sewing machines both in the United States and Europe, but it was another American who perfected their inventions and constructed a sewing machine which would use threads to form a stitch with the help of a needle.

The perfection of the sewing machine was to usher in the mass production of clothing. It ushered in an epoch cheap clothes for masses, more warmth, more cleanliness, more comfort.

The continued improvement of the sewing machine in recent years has reached a point where practically no hand work is required in garment making. Moreover, the electric machines used in modern factories today operate at a speed of 5,000 stitches per minute.

All parts of modern sewing machine are standardized so that it is very simple to get replacements for them. The standard of sewing machine made it easy to repair and to fix. Now more and more sewing machines are controlled by computers. Later sewing machines can be controlled by artificial intelligence. The sewing machines made by China have won worldwide attention and Butterfly has become a popular name. They can do many patterns we couldn't do in the past. We are sure the sewing machines will be better in the future.

New Words and Expressions

demand	要求	epoch	新纪元
manufacture	制造	warmth	暖和
industry revolution	工业革命	cleanliness	清洁
bring about	带来	comfort	舒适
serviceable	实用的，耐用的	improvement	改良
inventor	发明者	practically	实际上
perfect	完善的	moreover	再者，加之
construct	建造	electric	电的
form	形成	operate	运作
continuously	连续地	standardized	标准化的
perfection	完美	standard	标准
usher	引进，宣告	replacement	代替
intelligence	智能	worldwide	全球的
pattern	花型	artificial	人工的
mass production	大量生产	butterfly	蝴蝶

Notes to the Reading Practice 2

1. Like most important inventions, the sewing machine was the result of an important need of the time. 跟大多数发明一样，缝纫机也是时代需求的重要结果。
2. The demand for more speed and increased production in the manufacture of garments in the early industry revolution brought about the change of stitching. 对速度的更快要求和工业革命早期服装生产的增加带来了针迹改变。
3. The perfection of the sewing machine was to usher in the mass production of clothing. 缝纫机的完善推动了服装的大批量生产。
4. The continued improvement of the sewing machine in recent years has reached a point where practically no hand work is required in garment making. 最近几年缝纫机的不断改进使服装生产达到了不需要手工的程度。
5. All parts of modern sewing machine are standardized so that it is very simple to get replacements for them. 现代缝纫机的所有零件实现了标准化，因此得到配件很容易。

Lesson Twenty-Two

Text 1

>> Making Patterns for Trousers
制作裤子样板

A: Mr. Li, I have learned how to take measurements. What are we going to do next?

B: I'll teach you how to draw patterns, now let's draw a pair of trousers. We first draw front piece then back piece and in the end we draw other parts of the trousers such as waist belt, facing, pocketing and so on. Now let's draw front piece of men's trousers. On the supposition that the length is 104 cm, the waist is 76 cm, the hip is 100 cm, the rise is 30 cm, the bottom is 23 cm. Are you ready?

A: Yes, I am.

B: Now let's draw the first line. It's called basic line. The line is very important. It must be parallel with the selvage of the fabric. Next we draw the top line and bottom line. These two lines must be right angle with the basic line, and be parallel with each other. The bottom line is 100 cm from top line. Because the length of the trousers is 104 cm, we should reduce the waist belt, so the size is 100 cm.

A: What shall I do next?

B: Next we draw the hip line. It is 26 cm from top line. The rise size 30 cm minus waist belt 4 cm is 26 cm. Do you catch what I said?

A: Yes, I do.

B: Next we draw the knee line. It is 2 cm upwards from the middle point between hip line and bottom line.

A: Up to now we have drawn five lines. They are basic line, top line, bottom line, hip line and knee line. Am I right?

B: Yes, you are right. Now we draw hip width. It is 24 cm (one-fourth of the hip size minus 1 cm) from basic line. It must be parallel with basic line. Then we draw front crutch. It is 4 cm (4% of the hip size), next the waist width. It is 24 cm (one-fourth of waist size minus 1cm plus 6 cm).

A: Why must we subtract 1 cm from hip size and waist size?

B: Because back pieces of trousers are a little larger than front pieces. In order to meet the needs of body, back piece of trousers should be a little larger than front pieces. Oh, time is up. We have to stop here today. Let's have a rest for a change, shall we?

A: OK. Thank you. Good-bye, Mr. Li.

B: Good-bye.

New Words and Expressions

draw patterns	制图	top line	上平线
front piece	前片	up to now	到现在为止
back piece	后片	bottom line	下平线，底边线
in the end	最后	be right angle with	与……成直角
waist belt	腰带	be parallel with	与……平行
facing	门襟	minus	减
pocketing	袋布	hip line	臀围线
on the supposition that	假设，假如	width	宽度
length	长度	knee line	膝盖线
hip	臀围	upwards	上部
rise	直裆	front crutch	小裆宽
bottom	底边	plus	加
basic line	基本线	subtract	减
salvage	布边	for a change	换换脑子

Notes to the Text 1

1. On the supposition that the length is 104 cm, the waist is 76 cm, the hip is 100 cm, the rise is 30 cm, the bottom is 23 cm. 假定裤长是104厘米，腰围是76厘米，臀围是100厘米，直裆是30厘米，底边是23厘米。

2. It's called basic line. 这条边被叫作基本线。

3. Next we draw the top line and bottom line. These two lines must be right angle with the basic line, and be parallel with each other. 接下来我们画上平线和下平线。这两条线必须与基本线成直角并相互平行。

Text 2

>> Making an Inquiry 询价（询盘）

A: Excuse me. Are you Mr. Wang? I'm Peter.

B: Yes. I'm. Good afternoon, Peter. Nice to see you.

A: Nice to see you, too. We are interested in your dresses. I'd like to place a substantial order with you for silk dress. Please quote your most favorable price for silk dresses.

B: Thank you for being interested in our products. As you know, we are one of the largest exporters of silk dresses in our country and have been handling all kinds garments for more than fifties years. We are eager to enter into business relationships with you in the hope of developing trade between our two companies. We are pleased to give you a special offer, this is our most favorable offer. We offer you at $ 200 per dozen.

A: Oh, it sounds a bit too high. Your offer is too high to be acceptable. What about U.S. $ 150? I think the price is reasonable and acceptable to both of us.

B: We are not in a position to entertain business at your price. All our products are of top quality with moderated prices. If you have taken everything into consideration, you may find our price lower than those you can get elsewhere. But in order to conclude business with you at an early date, we can offer you a discount of 5%.

A: OK. Let's meet each other half way. In order to conclude the business, we can make some concession. We accept your price. We expect to have the Sales Contract drawn up in a day or two.

B: As required, we will draw up a Sales Contract for your approval. And we will discuss some details tomorrow and arrive at an agreement.

New Words and Expressions

make an inquiry	询价
a substantial order	大量订购
quote	报价
handle	经营
enter into	建立
a special offer	特别优惠报价
our most favorable offer	我们最优惠的报价
be in a position to do sth.	能做某事
entertain business	成交
moderate	适中的
take ...into consideration	将……考虑在内
conclude business	达成交易
meet each other half way	各让一步
make some concession	做些让步
draw up	草拟
as required	按照要求
approval	同意

Notes to the Text 2

1. All our products are of top quality with moderated prices. 我们的所有产品质量优良，价格适中。

2. meet each other half way 各让一步

3. In order to conclude the business, we can make some concession. We accept your price. 为了达成交易，我们可以做些让步。

Substitution Drills 替换练习

1. I have learned how to take measurements / draw patterns / design clothes .

2. Now let's draw a [pair of trousers / jacket / shirt / dress].

2. Now let's draw a	pair of trousers jacket shirt dress	.

3. It must be parallel with	the selvage of the fabric top line bottom line	.

4. These two lines must be right angle with	the basic line each other the top line the bottom line	.

>> Exercises to the Texts　课文练习 <<

1. 抄写并熟记本课短语和词组
2. 翻译下列短语
   1）on the supposition that
   2）hip width
   3）do sth. for a change
   4）take measurements
   5）top line
   6）be parallel with
   7）front crutch
   8）acceptable
   9）basic line
   10）be right angle with
   11）waist belt
   12）selvage of the fabric
   13）a little larger
   14）middle point
   15）facing
   16）bottom line
   17）a substantial order
   18）a special offer
   19）conclude business
   20）draw up
3. 将下列短语译成英语
   1）袋布
   2）与……平行
   3）与……成直角
   4）小裆宽
   5）做些让步
   6）各让一步
   7）询价
   8）成交

9）膝盖线
10）假设
11）基本线
12）直裆尺寸
13）上平线
14）腰围线
15）报价
16）减1公分
17）臀宽
18）臀围尺寸
19）裤子后片
20）比……稍大

4. 翻译

1）上平线必须与基本线成直角。
2）上平线应与下平线平行。
3）我教你如何绘制裤子的样板结构图。
4）为了符合人体后片应比前片大一点。
5）让我们听听音乐换换脑子。
6）到现在为止我们已缝制了5个领子。
7）接下来我们绘制裙子。
8）这条线被叫作上平线。
9）他们已经学会了画纸样。
10）这两条线必须互相垂直。
11）让我们互让一步吧。
12）他们的出价太高了，很难接受。
13）现在让我们绘制上衣的纸样。
14）这条线叫基本线，它很重要。
15）听懂我说的吗?
16）这两条线必须和上平线垂直。
17）我们必须把价格考虑进去。
18）这两条线必须互相平行。
19）我们不能接受你们的价格。
20）现在我们绘制裤子的纸样。

Conversation

>> Business Negotiation 商务谈判

A: Welcome to our company.

B: Thank you! I'd like to have a look at the clothes you are displayed. I see your products here. They look smart. It is said that your products are very popular with customers.

A: yes, you're right. You must have heard of our company. We are one of the largest clothing companies here and we have long history in producing clothes. These are our latest styles. Our products are novel in style, excellent in quality, advanced in workmanship and well sold at home and abroad. We trust these products of ours will appear to your market. Our products have met with the approval of clients all over the world. Our products are better in quality than those offered by other suppliers.

B: Yes. We think the superb workmanship and novel designs will appear to customers. Let me look at the quality of your products. We generally order after having seen the samples.

A: How do you feel like the quality of our products?

B: I find the quality is excellent; the products are suitable for our market. If the price is workable, we shall place a trial order with you. If the shipment proves satisfactory, we will place repeat orders with you for this product. What's the price of this coats?

A: The price of this coat is $ 60. Our goods are reasonable in price. We always sell goods at fair price. We have concluded considerable business with other clients at this price. In order to meet your request we will grant you a special discount of 5%.

B: Your price is agreeable to us. The price is acceptable to both parties.

A: Thank you! We are glad to have concluded this transaction with you. Owing to our joint efforts, our negotiation has come to a successful end.

New Words and Expressions

business negotiation	商业谈判	workable	可行的
look smart	看上去漂亮	well sold	畅销
novel in style	款式新颖	place a trial order	试定
appear to	迎合，投合	the shipment proves	供货令人满意
approval	认可	fair price	公平的价格
client	客户	meet your request	满足你们的要求
offer	出价	grant	答应
supplier	供应商	a special discount	特别的折扣
workmanship	工艺	both parties	双方
novel	新颖的	transaction	交易
sample	样品	owing to	由于

Reading Practice 1

Synthetic Fibers 合成纤维

Synthetic fibers can do many things that natural fibers cannot do. They are stronger than other cloth, and they wash better, or may hold their shape better. Synthetic fibers are used for warm sweaters that can be washed in a washing machine instead of having to be washed carefully by hand. They absorb little water, so they don't get dirty easily. Once they are dirty, they are washed easily. They are used to blankets, suits and carpets that will not be damaged by moths. They are used for ropes that lift great weights.

Most synthetic fibers have an interesting property—when they become hot, they soften. When are heated more, they melt. This property is useful because creases that are ironed into pants and skirts made these fibers last for a long time. On the other hand, care must be taken not to wash or iron synthetic fibers at too high a temperature.

Nylon is one of the most used synthetic fibers. It is strong and smooth and durable. Nylon fibers are also elastic. Pure nylon is used in women's stockings and underwear. It is used in ropes, fishing nets, and carpets. Natural fibers like wool are sometimes blended with nylon to make fabrics last longer. Wool socks, for instance, may have nylon added to the toes and heels, which get most of the wear.

New Words and Expressions

sweater	毛衣
blanket	毯子
carpet	地毯
rope	绳
lift	提起
property	特点
damage	损坏
moth	蛾
blend with	混纺
elastic	有弹性的
underwear	内衣
heel	足跟
soften	使软化
crease	折缝
durable	耐用的
stockings	长筒袜
toes	脚趾
wear	磨损

Reading Practice ❷

>> Some Important Steps of Document Operation 跟单的若干重要环节

As a department operator, our job is to carry out the contract smoothly. There are many steps in document operation. But some of them are essential. We must keep them in mind and make them clear to the departments concerned. It is clear that we should take the responsibility for the fulfillment of the order. We must do our best to execute our contracts to the full. We hope to cooperate with all the departments concerned and hope every department concerned to carry out the plan in time as contracted. We all hope, by our efforts, the goods are up to specifications as stipulated in the contract.

Inspecting the products on the assembling line is one of the most important tasks of a document operator. In the process of follow-up, quality assurance is the first priority. Quality control and inspection of products on the assembling line can guarantee the quality be the same as the sample. A careful inspection should be made in the process of the production. We document operators should make inspection of every process of the assembling line. There will be more frequent rounds of inspection by our personnel in the production of the goods.

We have to ensure that the products meet the quality requirements. A thorough inspection should be made before shipment. We require the quality be equal to the sample.

Delivering goods on schedule is one of another important facts we must ensure. We must realize that the time of delivery is a matter of great importance to us. Reputation is king. Reputation is our life.

>> New Words and Expressions <<

make clear	把……表达清楚	fulfillment	完成
concerned	有关的	execute	执行，实施
take the responsibility for	负责	as contracted	按照合同规定的

be up to	达到	frequent rounds of	好几轮
specification	规格	personnel	本人亲自
stipulated	规定的	thorough	彻底
inspect	检查（动词）	shipment	装运
inspection	检查（名词）	on schedule	规定的时间里
guarantee	保证	reputation	信誉

Notes to the Reading Practice 2

1. We must do our best to execute our contracts to the full. 我们必须尽力全面执行合同。

2. In the process of follow-up, quality assurance is the first priority. 在跟单的过程中，保证产品的质量是第一位的。

3. We all hope, by our efforts, the goods are up to specifications as stipulated in the contract. 我们希望通过我们的努力使货物达到合同规定的规格。

4. We must realize that the time of delivery is a matter of great importance to us. 我们必须认识到交货时间对我们来说是非常重要的。

Lesson Twenty-Three

Text 1

>> The Importance of Ironing and Pressing
整烫和熨烫的重要性

Mr. Li is an engineer in a garment factory. Today he is invited to tell us something about ironing and pressing. Here is Mr. Li. Let's have a welcome.

Ladies and gentlemen, I have the honor to give you a speech here about ironing and pressing. Ironing is an important task in tailoring and dressmaking. The fit and outline, the firmness and hang well of a garment depend on the ironing and pressing, so we should pay special attention to it when working. A good worker not only has skill in sewing but also knows how to iron and press a garment well. Nowadays we usually use pressing machines and steam irons instead of regular irons to iron or press garments. Pressing machines and steam irons are light and functional. They work well. They are used to press and iron cotton, linen, woolen, nylon and other materials. They are especially suitable for the final pressing of garments since they create steam. When pressing heavy or dark materials the damping cloth is needed to obtain a flat finish. We can choose one meter of mull, sheeting or tailor's canvas from which all the dressing has washed out. Some suppliers of water must be treated so that they can be used.

The other important thing is that once you enter the workshop, you should follow the teachers' instructions and observe the rules of the workshop. In the process of production safety assurance is the first priority. When you are pressing garments, you must always put "safety first" in mind, because a moment of carelessness will bring about a serious accident. Even if you leave your post for a short time, you must turn off the iron. Make sure that you turn off the iron before you leave the workshop after work. After work you must store your irons in an iron box. You should use and look after your irons well as the instructions tell you.

I tell you this so that you might do the work better. Wish you success.

That's all. Thank you.

New Words and Expressions

have the honour to do	荣幸地做某事	dressing	浆水
hang well	垂悬性好	observe	遵守
functional	实用的	instruction	教导
work well	好使	treat	处理
obtain	得到	supplier	供应商
flat finish	平整的效果	priority	最重要的
create	产生	process	过程
mull	细软布	firmness	坚固
sheeting	细布		

Notes to the Text 1

1. The fit and outline, the firmness and hang well of a garment depend on the ironing and pressing, so we should pay special attention to it when working. 服装的合体性、外形、挺括性、垂悬性都取决于整烫和熨烫，因此我们在熨烫时应特别注意。

2. Pressing machines and steam irons are light and functional. 熨烫机和蒸汽熨斗轻巧实用。

3. When pressing heavy or dark materials the damping cloth is needed to obtain a flat finish. 在熨烫深色或厚料的时候为了取得平整的效果要用湿烫布。

Text 2

>> Some Common Synthetic Fabrics 一些常见合成面料

Synthetic fabrics are in common use in garments. Synthetic fabrics are made of synthetic fibers and man-made fibers. There are three main types of synthetic fibers. They are true, regenerated synthetic and mineral fibers.

True synthetic fibers are made from materials such as coal, gas and petroleum. There are many types of fibers in synthetic fibers. They are nylon, dacron, polyvinyl fiber, polyamide fiber and so on, in which nylon and dacron are the most widely used synthetic fibers. Regenerated fibers are made from cotton, wood, straw and other materials, and mineral fibers from glass and asbestos.

Synthetic fabrics are stronger than natural fabrics. They are used to make up into trousers, jackets, work clothes, overall and so on. They wash better and keep their shape well after washing. They absorb very little water, so they dry quickly. But there are many weaknesses in synthetic fabrics. When you are pressing some sections and parts, you should make sure that the sizes and positions are right. If you press something wrong, it is difficult to do them well. It is very important to be careful when you iron and press the synthetic fabrics. When you are stitching or sewing some sections and parts, be sure to do them well, if you finish the garment and find something wrong, when you remove the stitches, the holes of stitches will be left in the materials or cloth. And it is the low absorption of water that makes people feel uncomfortable for clothing. They melt and scorch when they are ironed at high temperature, so steps are taken to improve their characters.

>> New Words and Expressions <<

main	主要的	weakness	弱点
regenerated	再生的	chlorofibre	氯纶
mineral	矿物的	straw	稻草
coal	煤	asbestos	石棉
petroleum	石油	section	部分

overalls	工作服	uncomfortable	不舒服的
shortcoming	缺点	melt	熔化
dacron	涤纶	scorch	烧焦
polyvinyl	聚氯乙烯化合物	absorption	吸收
polyamide	聚酰胺	character	特性

Notes to the Text 2

1. True synthetic fibers are made from material such as coal, gas and petroleum. 合成纤维由煤、天然气和石油等原料制成。

2. They wash better and keep their shape well after washing. 它们易洗且洗后成型性好。

3. They absorb very little water, so they dry quickly. But there are many weaknesses in synthetic fabrics. 合成织物吸水少，因而干得快，但是它们也有很多缺点。

4. They melt and scorch when they are ironed at high temperature, so steps are made to improve their characters. 当它们在高温熨烫时会融化烫焦所以要采取措施改善它们的特性。

Substitution Drills 替换练习

1. Today he is invited to tell us something about [pressing / designing / sewing].

2. I have the honor to give you a speech about [ironing / pressing / making patterns].

3. Once you enter the workshop, you should follow the [teachers' / workers' / coaches'] instructions and observe the rules of the workshop.

4. You should use and look after your [irons / sewing machines / tools] well as the instructions tell you.

5. He looks smarter and more energetic in a [T-shirt / jacket / suit].

Exercises to the Texts 课文练习

1. 抄写并熟记本课短语和词组
2. 翻译下列短语
   1）final pressing
   2）have the honor to do sth.
   3）wish sb. success
   4）pay special attention to
   5）hang well
   6）work well
   7）flat finish
   8）be invited to do sth.
   9）tell sb. sth.
   10）bring about
   11）be used to do
   12）sheeting
   13）melt
   14）turn off
   15）look after the iron
   16）heavy material
   17）create steam
   18）take steps to do sth
   19）observe the rules
3. 将下列短语译成英语
   1）湿烫布
   2）合体的外形
   3）洗掉
   4）铁箱
   5）服装的合体
   6）轻巧实用
   7）深色布料
   8）确保
   9）进入车间
   10）安全第一
   11）听从某人的教导
   12）整烫的重要性
   13）厚料
   14）最后熨烫
   15）浆水
   16）关闭
   17）一时的疏忽
   18）严重的事故
   19）保养好
   20）下班后

4. 翻译

1）当你离开岗位时，应关闭熨斗的电源。
2）服装的合体和挺括很重要。
3）熨烫厚料时，要用湿烫布。
4）在熨烫时应特别注意服装的垂悬性。
5）挑选一块细布或平布做烫布。
6）今天我很荣幸给孩子们上课。
7）我们要特别注意最后的熨烫。
8）我们必须把安全第一牢记心中。
9）她穿这条裙子看上去更漂亮。
10）我们必须保养好缝纫机。
11）进入车间要听师傅的话。
12）我们必须特别注意服装要合体。
13）这台缝纫机轻巧实用。
14）当你在熨烫服装时要注意安全。
15）下班后务必关闭电源。
16）她被邀请给学生们作报告。
17）现在越来越多的人意识到熨烫的重要性。
18）熨烫深色布料要用湿烫布。
19）我们必须遵守车间安全规章。
20）这些机器很好使。

Conversation

>> Signing a Contract 签合同

A: After a few days' talks, we have conclusion the business. Today we are going to sign a contract.

B: We are keenly interested in cooperating with you in expanding business. The contract we are going to sign is the result of our efforts. I am looking forward to our further cooperation and relations.

A: Please keep us posted of development at your country. We assure you that any further inquiries from you will receive our attention.

B: Please try your best to execute this order as it will bring you further business.

A: We assure you that we shall soon have the goods prepared and make everything ready.

B: We are glad to see the contract. Is there anything wrong with it? Would you please to go over the contract again?

A: I have read the contract carefully. There is nothing wrong with it. Let's get down to business to sign the contract now. We'll have two originals, each in both Chinese and English. Both are equally effective.

B: Where shall I put my signature?

A: Here, on the last page.

New Words and Expressions

sign a contract	签合同	keep us posted of	不断把消息传给我们
conclusion the business	达成交易	assure you	向你们保证，使你们放心
cooperate with	和……合作	receive our attention	受到我们的关注
expand business	扩大贸易	get down to business	着手
execute the order	执行合同	original	正本
bring you further business	带给你们下一笔生意	equally effective	同等效力
have the goods prepared	把货物准备好	signature	签名
joint efforts	共同努力		

Reading Practice 1

Document Operation (1) 跟单（一）

A: I'm keenly interested in document operation. Could you tell me something about it?

B: Document operation is a very important post in a company. Its main task is to ensure the orders which the customer placed be carried out completely.

A: I see. A person who works as a member of the document operation is a document operator. Am I right?

B: Yes, you are right. A document operator is a very important staff member in a company. He urges the workers and staff members in every department to fulfill or carry out the customer's orders in business process.

A: He makes frequent contacts with every department of a company, doesn't he?

B: Yes, besides all departments of his own company he will keep contacts with many government departments and service departments. So it is very important for him to master the skill of communication with others.

A: Is it necessary for a document operator to learn foreign languages well?

B: Yes, it is. Apart from foreign language, a well-trained document operator should have much study of business knowledge, computer and other subjects concern with the fields.

A: I must study hard and master all these knowledge as soon as possible.

New Words and Expressions

document operation	跟单	process	过程
ensure	确保	frequent	常常的，频繁的
work as	做，从事	contact	接触
place an order	订货	communication	交流
carry out	完成	apart from	除了
completely	全部地	subject	科目
document operator	跟单员	concern with	有关
urge	催促	field	领域
fulfill	完成	carry out	实施，履行
master	掌握	well-trained	训练有素的
much study	造诣深		

Reading Practice 2

Document Operation (2) 跟单（二）

A: Last time you told me something about document operation. Do you know what departments are involved in document operation?

B: Document operation includes buying fabrics (sometimes customers provide fabrics), designing, making patterns, cutting out, sewing, pressing, packing and shipping. A document operator should participate in every process and ensure that every staff member and worker completes his task in finest quality. He or she should carry out the contract without any delay. He or she will participate in business process.

A: Do you know what courses are involved in business activities?

B: Business includes the production, buying and selling of goods and services. They are inquiries, offers, acceptance the offer, counter-offer, placing orders, conclusion of business, signing a contrast, packing, payment, shipment, insurance, complaints and claims.

A: No customer, no business. We should establish good friendship relations with customers and make a great effort to promote business. If you want to take up this job, you need the ability to communicate with others in English freely.

B: You are right.

A: We shall do our best to learn English and computer well. We shall master the skill of making up clothes and the knowledge of foreign trade. We must know the whole process of document operation as soon as possible. I'm sure I will become a well-trained document operator.

New Words and Expressions

involved	涉及
provide	提供
ship	船运
participate in	参加
contract	合同
delay	耽搁
course	课程
inquiry	询价
offer	报价
acceptance the offer	接受报价
counter-offer	还价
conclusion of business	达成交易
insurance	保险
complaints	投诉
claim	索赔
establish	建立
effort	努力
promote	促进
foreign trade	外贸

Lesson Twenty-Four

Text 1

>> Laying the Patterns on the Material 排板

Before cutting out garments we should lay the patterns on the material. We should be careful when we lay the patterns on the material. It is important to do the job well. We first pin the patterns together, try on, and adjust to fit. Then we can lay the patterns on the material. The general rules for laying the patterns on the material are as follows.

1. If the material is to be folded in half lengthwise, be sure that the selvedge edges are exactly together. Some inexpensive materials are rolled badly during manufacture and the threads become pulled slightly out of the true direction so that when the selvedges are together the torn ends are not level. When this happens, two people should each hold one end and pull the material sharply diagonally until the ends become level.

2. Study the material for nap. If there is a pile or one way pattern, all the pattern pieces must be placed the right way up. If there is no nap, pattern pieces may be turned upside down, they should fit in more economically this way.

3. If the material has stripes or checks, make certain that these stripes and

checks will match at the seams.

4. Pencil the grain line on the pattern and make sure it lies on a straight thread of the material. Measure in from the selvage for accuracy, unless this is done the garment will not hang correctly, nor wear well.

5. Lay all the large pieces first and then, if possible, fit the smaller pieces into the spaces left.

6. Look for the markings donating that pieces are placed on the fold of the material and place these edges absolutely on the fold, for if they are placed only 0.5 cm inside the fold the garment will be 1 cm too large as the material is double.

7. If patter pieces have to be cut out singly, to avoid wasting material, remember to reverse the pattern when cutting for the second time, so that left and right sides are cut. If the pattern is not symmetrical, that is the left and right sides are different, the pattern pieces must be placed the right way up on the right side of the material in order that they may be cut for the correct sides.

8. When laying the patterns on the material, we should first lay the main pattern pieces, such as front pieces, back pieces and sleeves, then lay the small pattern pieces.

New Words and Expressions

lengthwise	长度方向	pile	长绒毛
see that	确保	economically	经济地
manufacture	制造	pencil	用铅笔画
selvage	布边	measure in from	从……边量
inexpensive	廉价的	singly	单独地
reverse	使翻转	accuracy	精确
roll	卷	stripes	条子
slightly	稍微地	donate	指示
torn	撕开的	absolutely	绝对地
diagonally	对角地	main	主要的
study	仔细看	symmetrical	对称的
nap	绒毛		

Notes to the Text 1

1. If the material is to be folded in half lengthwise, be sure that selvage edges are

exactly together. 如果织物是纵向对折的话，注意要使布边完全对齐。

2. ...the threads become pulled slightly out of the true direction so that when the sledges are together the torn ends are not level. ……以至于当扯开的布边放一起时布料的直丝稍有不顺。

3. Study the material for nap. If there is a pile or one way pattern, all the pattern pieces must be placed the right way up. If there is no nap, pattern pieces may be turned upside down, they should fit in more economically this way. 观察织物绒面，如有绒毛或一顺图案，那么所有的样板必须按一个方向排列。如无绒毛，样板可颠倒排列，这样做更加合适和省料。

4. Look for the markings donating that pieces are placed on the fold of the material and place these edges absolutely on the fold, for if they are placed only 0.5 cm inside the fold the garment will be 1 cm too large as the material is double. 找出样板中表示要放在布料对折线上的记号，然后将对折边准确地排在布料的对折缝上，如果对折处仅仅缩进0.5cm，那么服装将会大出1cm，因为布料是双层的。

Text ❷

>> Promoting Products 促销商品

A: Mr. Smith, this is the import and export commodity trade fair. Let me show you around the trade fair. This is the Clothing Hall. Welcome to our stand. Our products are novel in style, excellent in quality, advanced in technology, and well sold at home and abroad. All the exhibits are our best and most popular products. Please have a look at our products.

B: Thank you. Your jackets are most attractive and very popular with our customers. it is said that your products are very popular with customers. Chinese garments are popular with the customers with its special flavor, especially some Chinese elements.

A: Yes. These new products are popular in many places and countries. Chinese elements enjoy wide popularity in the world market. The “Dragon” logo has a wide application. Many Chinese elements such as “Double Happiness, Phoenix” enjoy wide popularity in the world market. Many foreign customers like to wear clothes with a dragon logo. We are willing to do everything to satisfy our customers. They enjoy the fine workmanship and good quality of our clothes. Our products are worth promoting.

B: I wonder if you can give me more information about the jackets you are showing?

A: Yes, I'd be glad to help you. Here you are. You will find our information helpful for you. The jackets are widely known. Our jackets sell well in the world market. This is our new design. All of our products are of good quality. Our price is reasonable compared with those in the international market.

B: Your products are found to our satisfaction. We expect to place with you an order for 500 jackets. We are sure that a trial order will be beneficial to both parties to the promotion of understanding.

New Words and Expressions

promote	推广，促销	application	应用
commodity	商品	Double Happiness	双喜
stand	摊位	phoenix	凤凰
special	特有的	satisfy	满足
novel	设计新颖	worth	值得
excellent	优异的	promoting	推广
technology	技术工艺	wonder	想知道
flavor	味道，风味	compare with	与……相比
enjoy wide popularity	受到广泛欢迎	satisfaction	满意
dragon	龙	expect	期待
logo	标志	beneficial	有益的

Notes to the Text 2

1. Our products are worth promoting. 我们的产品值得推广。“be worth doing” 意思是“值得做”，例如：The jacket is well worth buying. 这件夹克值得买。The book is worth reading. 这本书值得读。

2. We are sure that a trial order will be beneficial to both parties to the promotion of understanding，我们相信试定对双方促进了解是有益的。

Substitution Drills 替换练习

1. What are you hoping / planning / intending to do after school?

1. What are you	hoping planning intending	to do after school?

2. The general rules for	laying the patterns on the material ironing and pressing sewing and stitching	are as follows.

3. Be sure that	the selvedge edges are exactly together the grain line must be checked the patterns must be checked	.

Exercises to the Texts 课文练习

1. 抄写并熟记本课短语和词组
2. 翻译下列短语
   1）the fold of the material
   2）the right way up
   3）selvage edges
   4）reverse the pattern
   5）see that
   6）cut out
   7）pin the patterns together
   8）be folded
   9）out of the true direction
   10）study the material for nap
   11）be not symmetrical
   12）main pattern piece
   13）stripes and checks
   14）in half length
   15）one way pattern
   16）compare with
   17）be beneficial to
   18）match at the seams
   19）measure in
   20）not hang correctly
3. 将下列短语译成英语
   1）排料
   2）一顺花型
   3）用铅笔画直丝缕符号
   4）在缝份处对齐
   5）剩下的空间
   6）正面朝上

7）单片裁剪
8）不对称
9）寻找记号
10）从布边量
11）避免浪费
12）对称的
13）小片样板
14）做好这项工作
15）绒毛
16）条子
17）标有……的记号
18）确保
19）使翻转
20）双方

4. 翻译

1）裁剪前先排料。
2）我们的产品值得推广。
3）仔细检查样板上的直丝符号。
4）排料时，先排大片后排小片。
5）有绒毛的布料必须一顺排料。
6）无绒毛的布料可以颠倒排料。
7）排料的一般规则要牢记。
8）排料时要仔细。
9）我们的产品值得购买。
10）把布料铺好是很重要的。
11）铺料前应仔细查看织物的绒毛。
12）下周有一个进出口交易会。
13）我们的裙子很吸引人，非常受顾客的欢迎。
14）我们想向你们订购100条裙子。
15）一定要将布边对齐。
16）我们相信这次会谈对双方促进了解是有益的。
17）我们的裙子在世界市场销售得很好。
18）整烫和熨烫的总原则如下。
19）确保条子和格子的面料缝份处一定要对齐。
20）和市场上的同类商品相比我们的价格是合理的。

Conversation

Document Operation (3) 跟单（三）

A: Good morning, it's nice to see you. My name is Wangling. I'm from ABC Company. This is my card.

B: Good morning, welcome to our company! My name is Lipin. I'm delighted to meet you, too. I'd like you to meet Mr. Zhang, the head of our factory.

C: Good morning. I'm glad to see you. What can I do for you?

A: I'm a document operator from ABC Company. I'm mainly here today is to look at all the assembly lines, including cutting, sewing, pressing and packing. I'd like to see every procedure of the assembly line and the posts, the workers and staff members working. Would you please show me around your factory and give me some necessary introductions?

C: OK, I would be pleased to accompany you to our factory. This way, please.

A: Thank you Mr. Zhang, I'd like to look at the quality of your products. You know, quality is very important for both our two parties. The goods didn't sell well because the quality was poor last year. So we must pay special attention to the quality of the products.

C: You may rest assured of our close cooperation and we will do our best. Here we are. This is our cutting workshop. You see the workers are cutting out the pieces and parts of the clothes. Every piece and small part is laid according to their grain line. The stripes and checks are strictly matched at the seams. Now, we are in the sewing workshop. The workers are making their efforts to do their jobs well. Everyone is doing his best to execute this order.

A: Thank you for leading me to visit your factory and your production. I was so impressed by the high quality of your products. I really know more about your factory and production after this tour. I'm sure you will fulfill the contract ahead of time. Thank you for your cooperation.

New Words and Expressions

be delighted to do sth.	很高兴做某事	strictly	严格地
mainly	主要	execute this order	履行订单
procedure	步骤	tour	巡视
accompany	陪伴	be impressed by	留下深刻印象
rest assured of	请放心	ahead of time	提前
lay	排料		

Reading Practice 1

>> The Importance of the Markings
标记的重要性

There are two types of paper patterns, those which have instructions and lines printed on them and those which are perforated with various sized holes. Patterns are cut to the standard measurements and turnings are allowed outside these measurements. Most patterns have 1.2mm turnings allowed. The depth of the turnings is indicated on the pattern by printed line or by small perforated holes. This is the fitting line and it is of the utmost importance. It is, as its name implied, the line on which the garment fits. If the seams are up outside this line the garment becomes too large and if they are stitched inside the garment becomes too small.

Unless a style is not symmetrical one half the pattern is given, and when cutting out, some parts of the patterns may have to be placed to a fold in the material so that when opened out they are in one piece. When a part has to be so placed it may be shown on a pattern by three holes placed together to form a triangle.

The positions of darts, pockets and buttonholes are usually printed but may be marked with small holes and the latter are sometimes marked with perforations.

It is important to mark the grain lines on the patterns. This is very important to the fit and hang of the garment. So when we are making paper patterns make sure the grain lines must be marked.

>> New Words and Expressions <<

instruction	说明	symmetrical	对称的
print	印	form	形成
perforate	打洞	triangle	三角形
depth	深度	position	位置
indicate	指出	mark with	用……标出
utmost	最	perforation	孔
imply	含有	hang	垂势

Reading Practice ❷

Some Important Steps of Document Operation
跟单的若干重要事项

As a department operator, our job is to carry out the contract smoothly. There are many steps in document operation. But some of them are essential. We must keep them in mind and make them clear to the departments concerned. It is clear that we should take the responsibility for the fulfillment of the order. We must do our best to execute our contracts to the full. We hope to cooperate with all the departments concerned and hope every department concerned to carry out the plan in time as contracted. We all hope, by our efforts, the goods are up to specifications as stipulated in the contract.

Inspecting the products on the assembling line is one of the most important tasks of a document operator. Quality control and inspection of products on the assembling line can guarantee the quality be the same as the sample. A careful inspection should be made in the process of the production. Our document operators should make inspection of every process of the assembling line. There will be more frequent rounds of inspection by our personnel in the production of the goods.

We have to ensure that the products meet the quality requirements. A thorough inspection should be made before shipment. We require the quality be equal to the sample.

Delivering goods on schedule is one of another important facts we must ensure. We must realize that the time of delivery is a matter of great importance to us. Reputation is king. Reputation is our life.

New Words and Expressions

smoothly	顺利地	take the responsibility for	担起责任
make clear	解释	fulfillment	完成
concerned	相关的	execute	执行

cooperate	合作	guarantee	保证
carry out	执行	in the process of	在……的过程中
as contracted	按照合同规定的	requirement	要求
be up to	达到	frequent rounds of	好几轮
specification	规格	personnel	人员
assembling line	流水线	thorough	彻底的
stipulated	规定的	shipment	装运
inspect	检查（动词）	on schedule	规定的时间里
inspection	检查（名词）	reputation	信誉

附录一

拓展阅读 1：

Conversation

Tools for Sewing 缝纫工具

A: I'm interested in sewing and stitching. Could you please tell me something about tools for sewing and stitching?

B: Yes. Tool is very important. Sharp tools make good work. It's necessary to have right tools before beginning to cut out paper patterns and make up garments. There are many tools for us to use. I brought along some tools, they are scissors, needles, thimbles, tape-measures, all kinds of rulers and so on.

A: Is there anything else?

B: Yes, we need pencils and some tailor's chalks.

A: Where can we get these tools? I'd like your professional advice.

B: You can buy some of them from supermarkets. Some you may have to buy from specialist shops.

A: Why should we use these tools?

B: The professional result in sewing depends on right tools.

A: I see. Thank you.

2

A: Could you please tell me something about sewing machine?

B: Sure, the quickest way to sew is with a sewing machine. Sewing machine plays an important role in our job. Sewing machines are very high speed so they perform specific function in garment industry.

A: It is said that there are many kinds of sewing machines. What are they?

B: Generally speaking, we have lock stitch sewing machine which has straight stitching lock stitch. It is very useful for the sewing of most fabrics. So it is a must in sewing.

A: Besides lock stitch machines, have you got...?

B: We have zigzag machines, safety stitch over lock machines and buttonhole machines.

A: Thank you for your introduction.

B: Don't mention it.

New Words and Expressions

tools for sewing	缝纫工具
sharp tools make good work	工欲善其事，必先利其器
cut patterns	裁剪样板
make up garments	制作服装
bring along	带来
scissors	剪刀
tape measure	卷尺
tailor's chalk	划粉
specialist shop	专卖店
professional result in sewing	专业缝纫效果
depend on	依靠
professional advice	专业方面意见
perform specific function	起特别作用
industry	工业，行业
it is said	据说
generally speaking	一般来说
lock stitch sewing machine	锁针缝纫机
straight stitching	直缝
fabric	织物
must	必须的东西
zigzag machine	曲折缝缝纫机
safety stitch over lock machine	包缝机
buttonhole machine	锁纽孔机
don't mention it	别客气
at very high speed	高速

附录二

拓展阅读 2:

Conversation

Pressing 熨烫

A: Yesterday I attended a lecture at a vocational school. Mr. Li made a speech about pressing. He is an engineer of a garment factory.

B: Would you please tell me something about it?

A: All right. He told us that pressing is an important task in tailoring and dressmaking. Nowadays we usually use steam electric irons in garment factories.

B: What's the advantage of a steam electric iron?

A: Steam irons are light in weight but are useful for damp pressing.

B: What fabrics are they used to press?

A: They are used to press cottons, woolens, velvet, nylon and other materials. They are especially suitable for the final pressing of garments since they create steam.

B: Can we use steam irons to pressing heavy or dark materials?

A: Yes, we can. When we press heavy or dark materials, we need a damping cloth in order to obtain a flat finish.

B: What should we pay attention to when we press garments?

A: When you are pressing garments, you must always put "safety is the first" in mind. Pay close attention to safety rules and protect yourself from serious injury at work. When you are pressing garments, a moment of carelessness will bring about a serious accident. Even you leave your post a short time; you must turn off or disconnect the power plug. After work unplug the power cord from the outlet and store the irons in an iron box. I tell you the rules so that you never neglect it and might do the work better.

>> New Words and Expressions <<

press	熨烫	heavy	厚的
attend	参加	damping cloth	湿烫布
electric	电的	obtain	得到
advantage	优点	flat finish	平整的效果
light in weight	重量轻	safety	安全
damp	湿的	power plug	电源插头
fabric	织物	unplug	拔掉插头
woolens	毛料	cord	电线
velvet	丝绒	disconnect	切断
nylon	尼龙	outlet	出口
since	由于	store	放
create	产生	bring about	引起
dark	深色的	neglect	忽视
safety rule	安全条例	serious injury	严重伤害